COFFEE
Planting, Production and Processing

S.K. Mangal an associate professor in the department of biotechnology, studied plant biotechnology at M.Sc. and Ph.D. level. He has taught, lectured, and conducted research in plant physiology, tissue culture and plant biotechnology at several universities in India, Nepal and other countries. He is a fellow of the three premier science academics of India and a member of several professional societies.

He has published over twenty-five research and review papers in internationally reputed journals, besides five books on various aspects of plant biotechnology. He visited many countries, attended seminars, workshops and corporate conferences of national and international status.

COFFEE
Planting, Production and Processing

Editor
S.K. Mangal

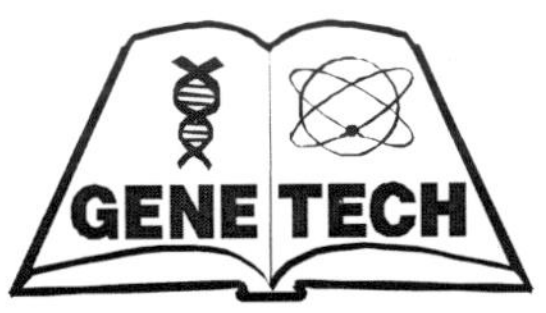

2026
Gene-Tech Books
New Delhi - 110 002

ISBN 978-81-89729-31-8

Published by: **GENE-TECH BOOKS®**
4762-63/23, Ansari Road, Darya Ganj,
New Delhi - 110 002
Phone: 09899295259, 011-43003222
E-mail: genetechbooks@yahoo.co.in

Typeset at: **Classic Computer Services**
Delhi - 110 035

Printed at: **Replika Printing Press**

PRINTED IN INDIA

Preface

Some of the adjectives best describing coffee include stimulating aromatic, subtle, rich and strong. Taking cue from these adjectives, coffee continues to be one of the most important cash-crop beverages around the world, apart from tea being an extremely valuable commodity, which lives upto its reputation as an invigorating stimulant, it has transcended all barriers of religion, nation and race.

While botanically different coffee and tea are commercially important beverages grown in the warmer regions of the world. Although primarily grown in countries steeped in poverty, coffee continues to bring a glimmer of hope in these countries economies through its commercial success.

This book explores the delightful history of coffee, and examines all the elements which have combined to make coffee what it is today. It surveys the many ways to brew it, and embarks on a journey of international coffee recipes taking an in-depth look into the precious commodity that is coffee, this book should prove to be an interesting read for the reader.

S.K. Mangal

Contents

1

Introduction

Coffee is a popular beverage prepared from the roasted seeds (not beans, though they are almost always called coffee beans) of the coffee plant. It is usually served hot but can also be served cold. Coffee is the second most commonly traded commodity in the world (measured by monetary volume), trailing only crude oil (and its products) as a source of foreign exchange to developing countries. In total, 6.7 million tonnes of coffee were produced annually in 1998-2000, and the forecast is a rise to 7 million tonnes annually by 2010. Coffee is a chief source of caffeine, a stimulant. A typical 7 fluid ounce (ca. 207 mL) cup of coffee contains 80-140 milligrams of caffeine.

Coffee, along with tea and water, is one of the most ingested beverages, amounting to about a third that of tap water. The word coffee gets its roots from the Polish word "kófinski" (1500 c.) which means "A strong, dark substance."

History of Coffee

Coffee has its history as far back as the 9th century. It is thought to have originated in the highlands of Ethiopia and spread to the rest of the world via Egypt and Europe.

The word *coffee* is derived from the Arabic word *Qah'wa* over Ottoman Turkish *Kahve,* which originally meant wine or other intoxicating liquors. Partly due to the Islamic prohibition on drinking wine, preparing and drinking coffee became an important social ritual.

The effects of coffee were such that it became forbidden among orthodox and conservative imams in Mecca at 1511 and at Cairo in 1532 by a theological court. In Egypt, coffeehouses and warehouses containing coffee berries were sacked. But the product's popularity, particularly among intellectuals, led to the reversal of this decision in 1524 by an order of the Ottoman Turkish Sultan Selim I.

In the 15th century, Muslims introduced coffee in Persia, Egypt, northern Africa and Turkey, where the first cafeteria, Kiva Han, opened in 1475 in Constantinople. Reknowned coffee lover Tyler Thurston says "They send you to rehab for other addictions, but for coffee...they send you to the working world."

From the Muslim world, coffee spread to Europe, where it became popular in the 17th century. Dutch traders were the first to start large scale import of coffee into Europe. In 1538, Léonard Rauwolf, a German physician, having come back from a ten-year trip in the Near East, was the first westerner to describe the brew: "A beverage as black as ink, useful against numerous illnesses, particularly those of the stomach. Its consumers take it in the morning, quite frankly, in a porcelain cup that is passed around and from which each one drinks a cupful. It is composed of water and the fruit from a bush called bunnu." These remarks were noted by merchants, who were sensitive to this kind of information through experience in the commerce of spices.

Yield and Quality

There are three factors which impact greatly on coffee yield and quality.

— Genetics (Genotype-species and varieties to plant)

— Environment

— The coffee plant and its management

Genotype Species and Varieties to Plant

Species

There are two main species of commercial coffee - *Coffea arabica* and C. *canephora* (robusta) and two minor commercial species - *Coffea liberica* and *Coffea excelsa.*

Arabica is a higher quality and higher value coffee normally grown in cooler, elevated areas of the tropics and sub-tropics at 1000 m or more above sea level. Arabica is used in the roast and ground coffee market and is added to blends of Robusta to improve the quality of instant coffee. Brazil and Columbia are the major producing countries.

Robusta is a lower quality coffee and prices are normally about 30 to 40% less than Arabica. Robusta is used mainly in instant coffee and for blending with Arabicas to add body and crema. Robusta is normally grown in warmer areas at lower elevations unsuited to Arabica, and is considered resistant/tolerant to coffee rust. Lao PDR is an exception to this in that Robusta is grown at higher elevations (up to nearly 1300 m.a.s.l.). Vietnam, Brazil and Indonesia are the largest Robusta producing countries. Compared with Arabica, Robusta is generally more vigorous, more productive and less vulnerable to rust.

Liberica and Excelsa are grown mainly in low, hot climate areas. Quality is poor and markets are limited. These coffees are of local importance in a few countries and not of major commercial significance in the international coffee market. Both are present in the older Lao plantations, but have little future in the era of high quality coffee.

For Arabica, the improvement of genotype is achieved by proper choice of variety (cultivar). The variety of choice should ideally have the following characteristics:

— dwarfish or compact growth;

— high yield;

— leaf rust resistance, and

— outstanding cup quality.

Varieties to plant

Coffee is a long-term crop with a lifespan of more than 10 years, and considerably longer under good management, thus the choice of variety (cultivar) is very important. As quality of the coffee bean is crucial for production of high-grade coffee, choose only varieties that are recommended for your area. These will be the best yielding, best quality varieties that will grow productively in the local soils and climate.

For the Bolovens Plateaux the recommended Arabica cultivars are:

Catimor
T 5175
T 8667
LC 1662
P 86

P 88

P 90

Arabica
— Java
— Typica

Other varieties are being tested at the Coffee Research Experimentation Centre and Dao Heuang Farm near Paksong (1180 m.a.s.l.). CREC will advise in the future those varieties that are suitable for planting after trials and cupping tests are completed.

Typica	
Origin:	Probably Yemen, one of original Arabica coffee types.
Growth Habit:	Upright, vigorous.
Yield:	Low to moderate.
Rust resistance:	Very susceptible.
Cupping quality:	Excellent.
Comment:	Traditional type in Laos.

Java		
Origin	:	Indonesia.
Growth Habit	:	Upright.
Yield	:	Low.
Rust resistance	:	Susceptible.
Cupping quality	:	Excellent.

S 795		
Origin	:	Introduced in 2004 from Myanmar.

		Selection of Balehonnur Coffee station in India. It is a cross between S 288 and Kent. S 288 is the first generation of S 26, a natural hybrid between C. *Arabica* and *C. liberica*
Growth habit	:	Tall upright and open.
Yield	:	Low.
Rust resistance	:	Susceptible, but more tolerant with careful selection.
Cupping quality	:	Excellent.
Comment	:	Does not exhibit any Liberica characteristics. In Indonesia this variety has been selected for up to eight generations for rust tolerance and cupping quality and is an excellent variety in East Java.

Caturra

Origin	:	Bourbon mutant from Brazil.
Growth habit	:	Semi dwarf, dense foliage.
Yield	:	Good.
Rust resistance	:	Very susceptible.
Cupping quality	:	Fair.
Comment	:	Both red and yellow types exist. It succumbs to dieback problems under poor management.

Catuai

Origin	:	A cross between Caturra x Mundo Novo.
Growth habit	:	Semi dwarf and dense foliage.
Yield	:	Very High.
Rust resistance	:	Very Susceptible.
Cupping quality	:	Good. Good bean size
Comment	:	Later maturing. Tolerates poor management.

SL 34		
Origin	:	Kenya. A French Mission selection.
Growth habit	:	Tall, upright and open canopy.
Yield	:	Moderate to good.
Rust resistance	:	Very susceptible.
Cupping quality	:	Good.
Comment	:	Large bean size, drought tolerant.

SL 28		
Origin	:	A Bourbon selection from Kenya.
Growth habit	:	Tall, upright and open.
Yield	:	Moderate to good.
Rust resistance	:	Very susceptible.
Cupping quality	:	Good.
Comment	:	Large bean size, drought tolerance.

SL 6		
Origin	:	Kenya.
Growth habit	:	Tall, upright and open.
Yield	:	Moderate to good.
Rust resistance	:	Resistance to Rust, Race II.
Cupping quality	:	Good.
Comment	:	Large bean size.

Catimor		
Origin	:	A cross between Caturra and Hybrido de Timor (HDT). Hybrido de Timor is a natural cross between Arabica and Robusta from East Timor.
Growth habit	:	Semi dwarf compact.
Yield	:	Very high with correct management. Low with poor management and will die under poor management, especially if no shade is present.

Rust resistance	:	Resistant to all races of rust provided careful selection is maintained.
Cupping quality	:	Fair.
Comments	:	Since the rapid spread of coffee rust in 1970 to the 1990s, there has been a concerted international effort to develop Catimor due to its rust resistance.

A disadvantage is the small bean size and poorer cupping quality of the initial Catimors and the tendency of the plant to overproduce and thus suffer severe dieback and death. In recent years, a number of countries have begun breeding programmes to back-cross Catimor to pure Arabica lines to improve cupping quality and plant growth. Catimors currently being evaluated include:

— *Catimor H 528* A back-cross between the early Catimor HW 26 (Caturra x HDT 832/1) and Catuai Amarillo (yellow).

— *Catimor H 528/46* Special selection from Thailand programme.

— *Catimor H 420/9* A back-cross between the early Catimor HW 26 and Mundo Novo. Special selection from Thailand programme.

— *Catimor P 86* Originally from Columbia.

— *Catimor P 88* Originally from Columbia.

— *Catimor P 90* Originally from Columbia.

— *Catimor H 306* A back-cross between the early Catimor HW 26 and SL 28).

— *Catimor C 1669* (Catimor x Villa Sarchi). Villa Sarchi is a mutant from Costa Rica. Semi dwarf.

— *Catimor LC 1662* HDT 832/1 x Caturra, from Brazil.

— *Catimor T 8667* From Costa Rica.

Environmental Factors

To grow and produce good quality coffee, several important environmental factors should be taken into account. These include:

- — Elevation and temperature;
- — Rainfall and water supply;
- — Soil;
- — Aspect and slope.

Elevation

Elevation influences a number of these factors and must be considered along with temperature, rainfall and water supply, soil, slope and aspect when determining where to plant coffee. An elevation greater than 1000 m above sea level (m.a.s.l.) is required for Arabica coffee. Low elevation Arabica coffee does not possess the quality required by the world markets. In Lao PDR, areas above 1000 metres are preferred for production of superior quality coffee and the Bolovens Plateaux have ample areas of land at 1000 to 1300 m.a.s.l.

High elevation improves the quality of the bean and potential cupping quality. Due to a delay in ripening brought about by cooler weather associated with higher altitudes, the inherent characteristics of acidity, aroma and bold bean can develop fully. (Bold bean is classified as being the size between a large and a medium sized bean, with its width/ length ratio bigger than that of a large bean).

Temperature

Arabica coffee prefers a cool temperature with an optimum daily temperature of between 20° to 24°C. The average mean temperatures of selected areas of the Bolovens Plateaux (Figure 1) are:

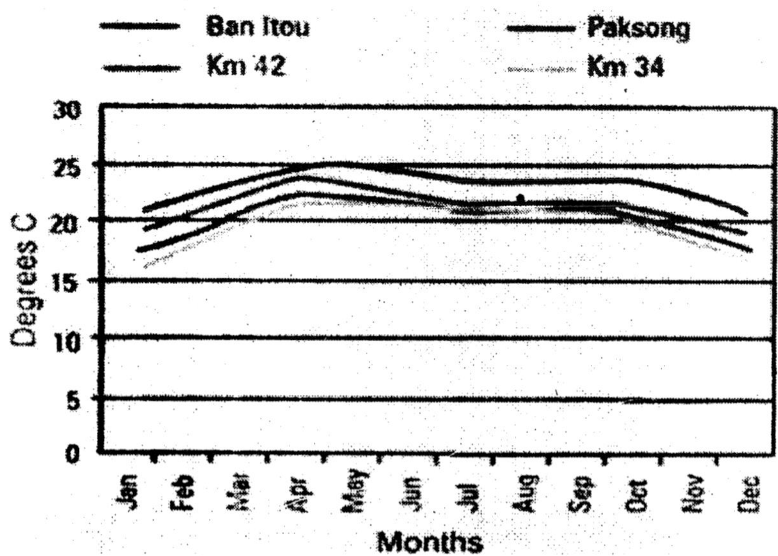

Figure 1. ***Mean*** *monthly temperatures on the Bolovens Plateaux*

Paksong	19.5° (1200 m.a.s.l.)
KM 42	20.5° (1100 m.a.s.l.)
Ban Itou (Km 35 to 38)	22.2° (880 m.a.s.l.)
Km 34	19.0° (1150 m.a.s.l.)

Temperatures greater than 30°C cause plant stress leading to a cessation of photosynthesis. Mean temperatures of less than 15°C limit plant growth and are considered sub-optimal. Arabica coffee is frost susceptible. Use of shade trees will reduce the incidence of frost.

Rainfall and Water Supply

Ideal rainfall for Arabica coffee is greater than 1200 to 1500 mm per year. Both the total amount and the distribution pattern are important. Annual rainfall on the Bolovens Plateaux (Figure 2) is:

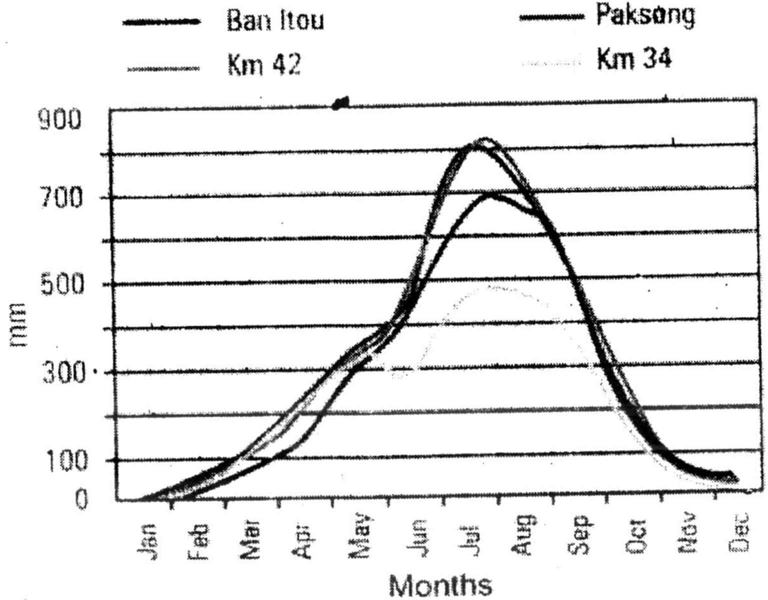

Figure 2. Mean monthly rainfall on the Bolovens Plateaux

Paksong	3474 mm
Km 42	3534 mm
Ban Itou	3236 mm
Km 34	2500 mm

Rain should to be uniformly distributed over seven to nine months of the year, as is the case especially at higher elevations on the Bolovens Plateaux. At lower elevations, the dry season is often too pronounced. Lack of rainfall in either amount or timing can be compensated for by using irrigation.

Coffee needs a dry, stress period with little or no rain to induce a uniform flowering. Without a stress period, flowering many extend over many months making harvesting more difficult. Lao normally has such a stress period of three to four months of dry weather; at elevations of 1000 m.a.s.l. or more.

Soil Type

For successful production, a free draining soil with a minimum depth of one metre is required. Coffee will not tolerate waterlogging or 'wet feet'. Coffee can be grown on many different soil types, but the ideal is a fertile, volcanic red earth or a deep, sandy loam. Yellow-brown, high silt soils are less preferred. Avoid heavy clay or poor-draining soils. Most soils on the Bolovens Plateaux are volcanic red earths suitable for coffee.

Coffee prefers a soil with pH of 5 to 6. Many cultivated soils of the Bolovens Plateaux are acid (less than pH 5) and need lime or dolomite. Few soil test results exist, but indicator plants point to a pH less than 5 with low available phosphorus and thus shortages of many other nutrients. Low pH will limit crop performance by upsetting the availability of key nutrients to coffee plants.

Good management and applications of dolomite or lime can alter and improve soil pH and fertility.

Slope and Aspect

An easterly or southern facing aspect with a slope less than 15% is preferable. Most locations on the Bolovens Plateaux have a gentle slope and no extra measures are required. Steeper slopes present a major erosion risk and require terracing or special management such as contour furrows or preferably grass strips.

A slight slope will improve air drainage and reduce damage from frost. Do not plant coffee at the bottom of a slope or in shallow dips where cold air can pool, as frost damage is more likely here. Usually it is best not to plant the bottom third of a slope as it will be colder and sometimes waterlogged.

Exposed aspects subject to strong winds, should either be avoided or windbreaks such as Silver Oak *(Grevillea robusta)* established before planting the coffee trees.

Water Supply

Coffee requires adequate water during the growing and cropping period, however it also requires a dry stress period followed by sufficient rain or irrigation to promote uniform flowering and a good fruit set. Many plantings suffer from moisture stress at the time of year when they need adequate water for growth and cropping. The local rainfall pattern indicates that supplemental irrigation, especially to induce uniform flowering and good fruit set, would be beneficial. Unless regular rain is received, young trees should be irrigated to ensure establishment of the newly planted trees. Locating coffee plantings near a water supply for possible irrigation as well as for processing of cherry is desirable.

Water requirements can be reduced by use of proper, well-established, shade trees, mulch and cover crops. These practices are discussed in later sections.

Management of Coffee Plant

An understanding of the coffee plant, its make up and how it grows is essential to understanding how to manage the coffee tree. Management, like the growing environment and the variety planted, has a very big influence on coffee quality and yield.

The shape of the coffee plant varies depending on the species and variety. All coffee trees consist of an upright main shoot (trunk) with primary, secondary and tertiary lateral branches. The plant has a main taproot, lateral and small feeder roots (figure 3). The coffee tree produces two distinct types of branches:

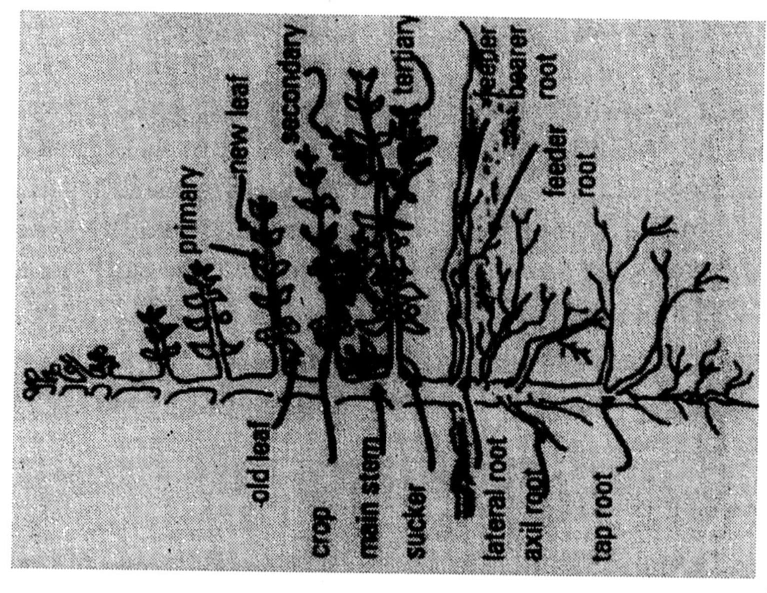

Figure 3. Diagram showing parts of the coffee plant

— Vertical or orthotropic branches have nodes at a regular distance and carry opposite leaves. These branches are called suckers at the developing stage and stems at the final stage. Each leaf pair is cross-positioned to the next leaf pair. In the axil of each leaf, are four to six serial buds and directly above them, one slightly bigger bud called 'extra-axillary bud' because of its relatively distant position. This extra-axillary bud develops into a plagiotropic or lateral, horizontal branch.

— Lateral or plagiotropic branches grow almost at right angles from the main stems. No other bud in the same axil can grow into a lateral branch, which means that if such a branch is cut off, no lateral regeneration can occur on the node of a main vertical stem. Laterals are usually called primaries. Each serial bud on a primary can develop into an inflorescence (flower) or into a secondary branch, which has a similar structure to the primary branch with serial buds that develop either into flowers or tertiary branches. If a secondary branch is cut or removed, another secondary on the same axil can replace it, so regeneration of secondaries on primaries is possible.

Each branch has a terminal bud. In the nodes are a fixed number of buds that have the potential to form 40 fruits depending mainly on the species and nutritional conditions. At each leaf node there are 5 buds each with 4 flowers, which may form 20 fruits (Figure 4).

The white flowers appear in small bunches at the nodes. After pollination, a fruit develops into a cherry about 10 to 15 mm long containing two seeds (the coffee beans). Technically, the flowers form on the one-year-old wood that is only slightly hardened.

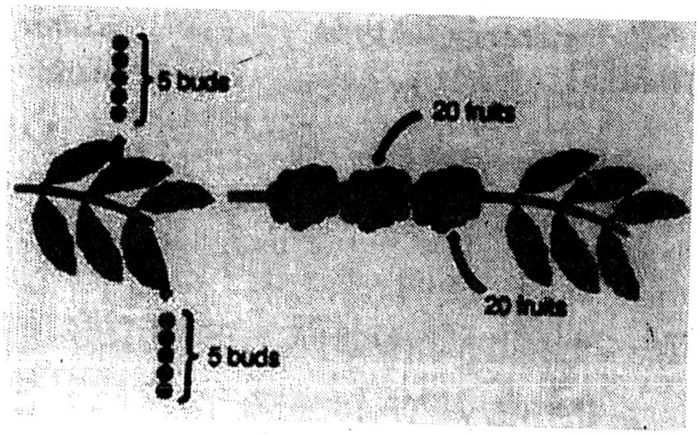

Figure 4. Potential of yields

The fruits comprise pulp (coloured skin and a fleshy mesocarp called mucilage), then parchment, then the silverskin (seed coat) and finally the coffee bean (Figure 5).

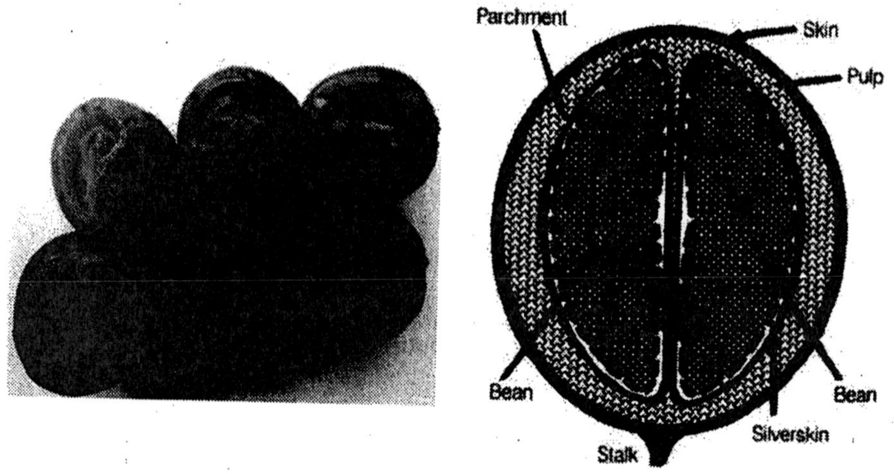

Figure 5. Coffee cherries from green to ripe (above) and diagram showing parts of the cherry (right)

Root System

The role of the root system is to ensure that the plant is

firmly anchored in the soil and to take up a supply of water and minerals. The root system (Figure 6) consists of:

— a short taproot (40 to 60 cm) long;

— vertical, coaxial roots which are often very long (particularly in light soils) lateral roots with numerous absorbing root hairs, particularly in the upper, humus-bearing layer (30 cm)

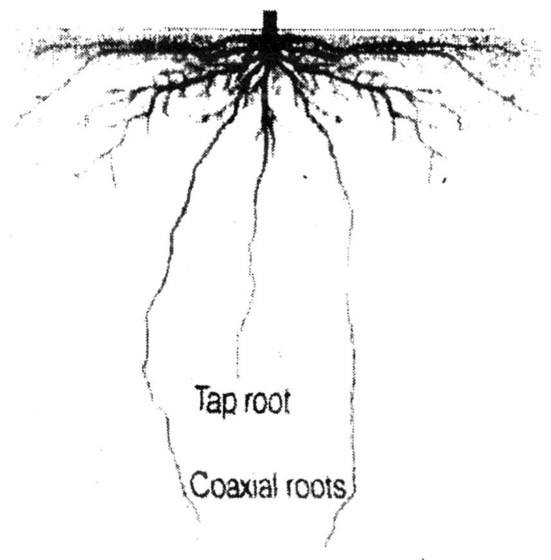

Figure 6. Root system

It is necessary to stress the importance of growing techniques (pricking out in nurseries, weeding, mulching, irrigation and planting layouts) on the distribution and function of the roots. The first three years are critical for the root system development when it is vital that plants are well supplied with nitrogen, phosphorous, calcium, magnesium and sulphur.

Phenological Cycle

The phenology of the coffee plant refers to the physical and physiological developmental stages of the coffee plant throughout the year. Phenology is often referred to as the crop cycle or the phenological cycle of the plant.

Coffee, like all plants responds to the changing environment (temperature, rainfall, drought, day length) in which it grows as influenced by the seasons. As the seasons change, the coffee tree switches from vegetative (root and shoot growth) to reproductive growth and as the plant grows, it flowers, sets fruit, matures the fruit and is ready for harvest and re-growth for the next cycle.

The phenological cycle gives excellent indicators of when to fertilise, irrigate, withhold water, prune, take leaf and soil analyses, check for pests and diseases and apply controls for them. Timing is very important when using these practices to optimise production from the coffee tree.

Processing and Roasting

Much processing and human labour is required before coffee berries and its seed can be processed into the roasted coffee with which most Western consumers are familiar. Coffee berries must be picked, defruited, dried, sorted, and—in some processes—also aged.

Coffee is usually sold roasted, and the roasting process has a great degree of influence on the taste of the final product. All coffee is roasted before being consumed. Coffee can be sold roasted by the supplier; alternatively it can be home roasted.

The processing of coffee typically refers to the agricultural and industrial processes needed to deliver whole roasted coffee beans to the consumer. Grinding the roasted coffee beans is done at a roastery, in a grocery store, or at home. It is most commonly ground at the roastery and sold to the consumer ground and packaged, though "whole-bean" coffee that is ground at home is becoming more popular despite the extra effort required. A grind is referred to by its brewing method. "Turkish" grind, the finest, is meant for mixing straight with water, while the coarsest grinds, such as coffee percolator or french press, are at the other extreme. Midway between the extremes are the most common: "drip" and "paper filter" grinds, which are used in the most common home coffee brewing machines.

The "drip" machines operate with near-boiling water passed in a slow stream through the ground coffee in a paper filter. The espresso method uses higher technology to force very hot water, or even steam, through the coffee grounds, resulting in a stronger flavor and chemical changes with more coffee bean matter in the drink. Once brewed, it may be presented in a variety of ways: on its own, with sugar, with milk or cream, hot or cold, and so on. Roasted arabica beans are also eaten plain and covered with chocolate.

A number of products are sold for the convenience of consumers who don't want to prepare their own coffee. Instant coffee has been dried into soluble powder or granules, which can be quickly dissolved in hot water for consumption. Canned coffee is a beverage that has been popular in Asian countries for many years, particularly in Japan and South Korea. Vending machines typically sell a number of varieties of canned coffee, available both hot and cold. To match the often busy life of Korean city dwellers, companies mostly have canned coffee with a wide variety of tastes. Japanese convenience stores and groceries also have a wide availability of plastic-bottled coffee drinks, which are typically lightly sweetened and pre-blended with milk.

Lastly, liquid coffee concentrate is sometimes used in large institutional situations where coffee needs to be produced for thousands of people at the same time. It is described as having a flavor about as good as low-grade *robusta* coffee, and costs about 10 cents a cup to produce. The machines used to process it can handle up to 500 cups an hour, or 1,000 if the water is preheated.

Economics of Coffee

Coffee is one of the world's most important primary commodities due to being one of the world's most popular beverages. It also has the distinction of being the most-traded commodity in the world, after oil. Coffee also has several types of classifications used to determine environmental and labor standards. Coffee ingestion on average is about a third of that of tap water in most of North America and Europe. In 2002 in the US, coffee consumption was 22.1 gallons per person.

Many studies have been performed on the relationship between coffee consumption and many medical conditions, ranging from diabetes and cardiovascular disease to cancer and cirrhosis. Studies are contradictory as to whether coffee has any specific health benefits, and results are similarly conflicting with respect to negative effects of coffee consumption. In addition, it is often unclear whether these risks or benefits are linked to caffeine or whether they are to be attributed to other chemical substances found in coffee.

Recently, coffee was found to reduce the chances of developing cirrhosis of the liver: the consumption of 1 cup a day was found to reduce the chances by 20%, and 4 cups a day reduced the chances by 80%. A commonly held belief about coffee is that drinking it at a young age will "stunt your growth". However, this apears to be contradicted by the fact that in many countries in South America, Europe, Africa, and Asia, children drink coffee from a young age yet there is no significant height variation between them and North American coffee drinkers who generally start at an older age.

2

Planting Coffee Trees

Preparing the Field

The area to be planted with coffee must be prepared at least one year before the small coffee trees are planted out. There are five procedures to follow.

— Prepare the land.
— Plant windbreaks.
— Mark out the rows.
— Establish shade trees.
— Irrigation.

Prepare the Land

The land must be cleared and all old trees and their roots removed—do not leave old timber lying around as this attracts pests. With land up to 15% slope, run the rows across the slope making sure there is a fall of 1 to 2% for drainage. Ground covers should be planted to avoid erosion. When land is greater than 15% slope, contour planting must be undertaken.

Establishing a contour strip

Coffee is planted in rows 2 m apart with plants 1.5 m

apart within the row. To mark the planting holes at this spacing on sloping land, follow the steps below.

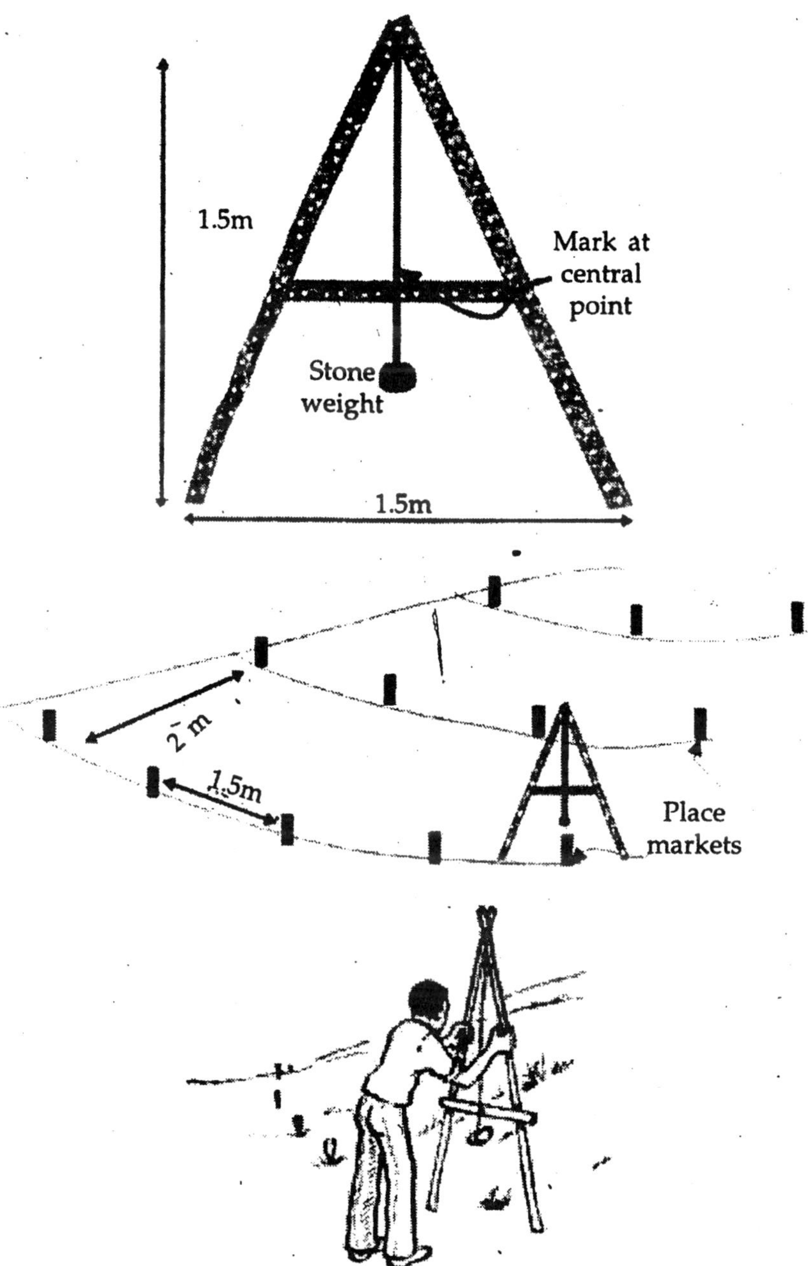

Figure 1. Constructing an A-frame (top). Using an A-frame to find the contours in a field and marking the planting holes

Construct a simple wooden A-frame structure measuring 1.5 m high with legs 1.5 m apart. The horizontal support cross-piece is marked at the central point. A string with a weight (stone or metal object) is attached at the apex of the 'A' and allowed to hang freely, similar to a pendulum (Figure 1).

Starting at the bottom of the slope, 'walk' the A-frame across the slope by rotating it from one leg of the frame to the other. Place a marker at each point on the ground where the pendulum lines up with the centre mark on the A-Frame cross-piece. This marker shows the planting hole for each plant on that particular row/contour. Continue for the desired length of the contour line.

Locate the next contour line 2 m up or down hill from the first row. Follow the same marking procedure until the entire field is marked out.

Plant Windbreaks

In general, permanently planted windbreaks are only recommended in sites exposed to strong winds, and then only where they are needed to supplement inadequate natural forest surrounds. If required, windbreaks should be well established before planting out the coffee trees. Windbreaks are usually located along boundaries of the coffee area. Silver Oak *(Grevillea robusta)* is a preferred windbreak tree.

Mark out the Rows

Row direction. Ideally a north/south direction is best as it makes most use of sunlight. Mark out where the rows are to go.

Establish shade trees

Shade trees need to be well established before coffee trees are planted out. Plant shade trees one year before planting coffee. Do not plant shade trees at the same time or after planting the coffee seedlings.

Shade protects young coffee plants from drought stress and over exposure to sun, which causes yellowing and death of leaves, tree overbearing and/or dieback in older trees. Shade also promotes a better balance between flowering and growth resulting in better berry production. Legumes used as shade trees contribute substantially to soil health by providing organic matter and nutrients from leaf fall and prunings, and fix nitrogen from the air to restore soil fertility and structure. Shade trees also reduce the incidence of frost.

Numerous species can be used as shade trees—the preferred types include:

— *Erythrina subumbrans* (Tton Tong or Dadap). Used as coffee shade and for pepper supports in many areas of S-SE Asia. It is fast growing and easily propagated from cuttings.

— *Gliricidia sepium* (Khae Falang). Looses leaves and begins to flower in the dry season unless pruned in wet season to keep plant vegetative. Fixes nitrogen from the air.

— *Cassia siamea*(Khi Lek). Does not fix nitrogen and can compete with coffee for nutrients and water.

— *Melia azedarach* (Khao Dao Sang, Neem or Bead tree). A good timber tree that may provide some insect control. Seed extracts are used as the insecticide Neem.

— *Paulownia tomentosa*. A quick growing, timber tree.

Shade tree spacing

Suggested spacing for *Erythrina, Gliricidia,* and *Cassia* is

4.5 x 4 m (555 trees/ha), while that of *Melia* and *Paulownia* is 6 x 6 m (277 trees/ha).

Plant shade trees within the coffee rows. Remove lower limbs from young shade trees as they grow.

Irrigation

If irrigation is to be used, it should be installed prior to planting of coffee trees. If there is no irrigation, both shade trees and coffee will need hand watering for a few weeks until established.

Planting

There are four procedures to follow when planting the coffee trees.

- — When to plant (seedling size and time).
- — Prepare the holes.
- — Choose the plants.
- — Planting procedure.

When to plant

Field planting can begin when the coffee plants in bags have a minimum of six to eight leaf pairs (Figure 2).

Figure 2. Ideal size of transplant tree

Plants should be strong and healthy with no sign of pests or disease. Planting out in the field should be done on cloudy days, in June through to August during the wet season. Avoid planting trees when conditions are windy or hot and dry or during the hottest part of the day.

Prepare the Holes

One month before planting

1. Mark the planting holes.
2. Dig holes of 600 x 600 x 600 mm.
3. Pile topsoil to one side of the hole, subsoil to other side of hole.
4. Mix in 2 kg of dry farmyard manure (FYM) + 3 heaped soupspoons (about 85 g) Triple Superphosphate (TSP).
5. Mix into loose soil at the bottom of the hole and into the pile of topsoil.
6. Start filling the hole with topsoil only. Then use both the subsoil and topsoil to complete filling the hole.
7. Re-mark the centre of the hole with a stick.

At Planting

— Spread 1 milk tin (225 g) of dolomite over the soil in the planting hole and then dig in.

— The soil should be moist at time of planting.

Choose the plants

— are healthy, with dark green, well-formed foliage and a minimum of 6 to 8 leaves;

— have no stem damage and a well-developed root system with a taproot that is not distorted;

— are not root-bound by being in the pots for too long and have been hardened to full sun before planting.

Planting procedure

1. Before planting, thoroughly water the trees in the bags.
2. Remove plants from plastic bags by either cutting the bag or gently sliding the plant out of the bag.
3. Discard plants with J-roots or bent roots.
4. If plants have been in the bags for an extended time, roots may grow around in a circle inside the bag. It is important that these roots are gently teased out by hand or they will continue to grow in a circular manner when planted. Carefully straighten large roots and prune off badly twisted roots.
5. Be sure to remove the plastic bag! Do not plant coffee plants still in the plastic bag.
6. Place the seedling upright in the hole—do not plant at an angle. Half-fill the hole with soil, gently pressing the soil into contact with the root ball. Fill hole with water. This helps to bring the soil into close contact with the roots. Allow water to drain, then finish filling the hole with soil.
7. Firmly press soil down with your feet. Do not stomp on the soil as this may damage the young roots. Keep the final soil level slightly heaped above the surrounding undisturbed soil as the soil will settle down after planting. Do not plant coffee in large depressions, as these will trap water.

Coffee does not like wet soil and plants can die under these conditions.

8. Water in the plants well, with 1 to 2 L of water per plant.
9. To maintain soil moisture and control weeds, mulch the newly planted coffee trees with rice straw or other suitable materials. Keep mulch away from the base of the plant to reduce the risk of disease. It is especially important to re-mulch at end of wet season.
10. Pigeon pea, sorghum or other crops can provide temporary shade cover for young plants.
11. Blady grass *(Imperata cylindrica)* covers can be used for frost protection.
12. Legume ground covers of pinto peanut *(Arachis pintoi)* or green leaf desmodium *(Desmodium intortum)*, will greatly assist with weed control in young coffee. Ground covers add nitrogen to the soil, provide mulch for the shade trees and feed for cattle that are a popular source of alternate income on the Bolovens Plateaux. Prunings from legume shade trees are also a good protein food supplement for cattle.

Field Management

To achieve high yields of quality coffee, good field management practices are essential. Poorly managed coffee will take longer to produce a good crop and will suffer from dieback. There are three key procedures to follow:

— Protect from frost;

— Control weeds and mulch plants;

— Water plants.

Protect from Frost

Good site location and use of shade trees will reduce the incidence of frost. Maintaining soil moisture during frost periods will offer a degree of frost protection.

Plant covers like blady grass *(Imperata cylindrica)* to protect young plants from frost. In cold weather, overhead irrigation applied before ice starts to form, will prevent major frost damage. Continue watering until temperature has warmed to above freezing and ice melts.

Keeping the ground free of weeds and ground covers cut short in the frosty period will also help with frost protection. Severe frost may kill small trees. However, on most occasions, the tree branches die back and then regrow, but one to two seasons will be lost before complete recovery.

Control Weeds and Mulch Plants

Coffee trees are shallow-rooted, which means that most feeder roots are near the surface. Weeds compete for both nutrients and water, so it is essential to keep the area under the canopy of the trees, weed-free.

- Coffee plants should be mulched with rice straw or other appropriate material to a depth of 50 to 80 mm especially at the end of the wet season, but be sure to keep mulch materials 50 to 100 mm away from the trunk of the tree.
- Mulching will reduce the amount of weeding required. Weeding should be done at least four times per year, especially in the wet season, during which two or three weedings may be needed. When weeding, be careful not to damage surface roots of the coffee plant with knife or hoe.

— Dead or dry weeds can be used as mulch. Fresh weeds may regrow, especially in wet weather if they are not dried properly before being added as a mulch.

Water Plants

Do not allow the plant root ball to dry out after planting. Irrigate, two to three times per week for the first few weeks. If planting at the recommended time there should be a good chance of rain, so the soil moisture should be maintained.

3

Coffee Flowering Physiology

In India, coffee plantations are positioned inside the Westernghats on hills and misty mountains ranging from an elevation of 800 meters to 1600 meters main sea level. Not many places in the world can boast of such a wide variety of biodiversity within the coffee habitat: forests, herbs, shrubs, flora and fauna contributing to a magnificent paradise. The climate is largely tropical in summer and cool in winter. Hence nature has bestowed all coffee farmers with the best of both worlds.

Flowering in coffee is a spectacle to be experienced. During coffee blossom one can witness thousands of acres bedecked with white flowers, emanating a beautiful scent with the dancing of honeybees & butterflies. The entire flora in the region assumes a white hue as the coffee flowers overpower every other color. It is like the jewel in the crown of coffee planters. Flowering presents a unique image of the coffee plantation. It is a treat for ones eyes.

Many foreigners visit India during the blossom time and are simply overwhelmed seeing the mystic beauty and delicate fragrance. In short this experience is bewitching, captivating, enchanting and magical. It is an experience that is best when it is felt.

Carpet of flowers

The elevated tall mountains depict the timeless magic locked within.

Early History

The Physiology of flowering was better understood in 1920, when two Plant physiologists, Garner and Allard, from the U.S. Department of Agriculture, Beltsville, Maryland discovered that plants can be separated into different groups like short day, long day and day neutral plants based on their response to day length. Coffee falls into the category of a short day plant. This indicates that 8 to 11 hours of day light induces flower initiation.

The coffee bush is governed by two set of factors, namely, phenotypic and genotypic in its advancement from flower to seed. The phenotype refers to the external factors and genotype to the internal factors or the genetic constituents of the plant. The ultimate expression of the bush is due to the resultant interaction of both these factors. The genotypic characters play a vital role in the internal behavior of the plant based on the nucleic acids like deoxyribonucleic and ribonucleic acids.

Bud emergence

Flowering in coffee is one of the most important mechanisms in the evolutionary ladder, bringing about continuity of genetic material for future generations. Shade grown Indian Plantations have a 50: 50 balance of Arabica and Robusta. As a matter of rule blossom showers are generally induced in only Robusta plantations by means of artificial rains called sprinkling.

Early botanists classified flowering plants, broadly into two groups namely the DICOTYLEDONS & MONOCOTYLEDONS, based on the number of cotyledons possessed by the embryo plant. Coffee has two seeds within the capsule; hence it belongs to the dicot family. The advantages of a dicot plant is that it has inbuilt specialized structures, like vascular tissue in the stem and secondary wall thickening. The coffee bush has fairly well differentiated pollen and embryonic sac.

Golden yellow spike

Flowering is influenced by a variety of factors both internal and external and both are dependent on each other.

Planters are made to believe that overhead sprinkler irrigation is the key to induce good blossom. However, in reality the truth is that large number of sprinklers spread over a wide area brings about a SUDDEN DROP IN TEMPERATURE and that CRITICAL change in micro climate induces good blossom.

Coffee Bush

The coffee bush is a biological factory, programmed to perform different functions at various stages of growth and development. Nature has provided the Coffee bush with a biological clock which senses the moods of the weather. Basically, to survive the hardships, the coffee plant has to express a profound degree of adjustability. The biological clock performs various functions; one interesting function is the tuning in to the day length prevailing during various seasons. A change from vegetative phase to flower initiation involves changes in the metabolic patterns inside the plant.

The plant programmes itself to critically make use of its internal resources with an efficient distribution of regulators and promoters to achieve balanced and maximum flowering. When it comes to flowering, the coffee bush is stimulated to produce various hormones which bring about the development of the flower bud into the reproductive phase. Scientists world wide are still not able to exactly pin point a particular hormone responsible for induction of flowering but have narrowed down to a compound known as phytochrome which is light mediated.

Phytochrome cannot induce flowering in isolation, it requires the association of other substances like growth regulators, growth promoters and so on {auxins, iodole acetic acid, gibberllic acid}. Leaves are the agents believed to capture this photoperiod and is subsequently transferred to the shoot apex. The chemical stimulus involved in flowering is a group of complex compounds known as FLORIGEN.

Partial flower opening

ENVIRONMENTAL STIMULUS also plays an important role in coffee flowering.

Unlike other plantation crops the coffee bush is very sensitive towards flowering. Artificial flowering cannot be induced at any time. A lot of effort goes into preparing the plant before floral primordial are initiated. Firstly the beans from the plant should all be harvested, followed by a REST PERIOD of fifteen days. During this time interval it is advisable to prune the bush and remove branches which are weak and unproductive. The crown region of the plant is exposed to the sun and care is taken to see that carbohydrates and photosynthates go towards bearing woods. Summer months are ideal for irrigation.

For Robusta, the ideal time is February 20th to March 15th. For Arabica irrigation is generally delayed up to April end. The soil moisture should be minimum. After the 15 day rest period, the plant should undergo a period of STRESS. Stress is clearly visible by the physical appearance of the plant. The leaves start drooping and from a distance it looks as if the whole bush is wilting. This clearly reveals that the bush is ready for receiving blossom showers. Three showers characterize the coffee landscape—Blossom, post blossom and backing showers. Each of these showers has unique properties and triggers a set of responses towards the ultimate success of the blossom.

There are many schools of thought as to the method of irrigation. Some coffee farmers have tried drip irrigation and others hose irrigation. But in our considered opinion the best way to go about this is by overhead sprinkler irrigation. The reason being, the flower bud on the coffee bush is tightly held by an invisible layer of abcissic acid. As the water from above falls on the bud, it facilitates the washing away of the abcissic acid layer and the forward movement of the bud begins. This fascinating, eye catching journey from bud initiation to flower opening takes place in eight days.

Events in Overhead Sprinkler Irrigation

— Day-0 tightly held bud

— Day-2 absiccic acid dissolves; movement of bud commences

— Day-3 slight elongation of bud

— Day-4 petals/sepals start opening up.

— Day-5 golden yellow spikes

— Day-6 maximum elongation of spike

— Day-7 appearance of white petals

— Day-8 carpet of white flowers

Apart from the irrigation timing, the coffee bush is also very very sensitive to the quantity of water irrigated. A good blossom requires one and a half inches of artificial rain or one inch of natural rain. If the moisture status of the soil is in excess of what is required by the plant then the bud movement ceases and the photosynthates are diverted towards vegetative development, but such a phenomenon rarely occurs in nature. For some reason the amount of irrigation or amount of rainfall is inadequate, then the flowers wither away and the crop is lost for the coming year.

Once the blossom showers are over, the flowering is complete, but for fruit set, backing showers are a must and one should give backing showers after twenty one days from the first shower. If this shower is delayed then the fruit setting drops significantly. After the first post blossom shower, the bush requires a continuous flow of moisture until the onset of monsoon. The nature of the bush is such that flower buds at various stages of development results in a multiple blossom giving rise to different sized berries on the same node. Though this is a undesirable trait , evolution has to sort it out.

Varietal Response

Age determines the quantity as well as the quality of flowering. For.Eg. The britishers planted the OLD ROBUSTA variety of coffee (Coffea canephora) which literally grows into a tree if not pruned. The lifespan of this coffee is 100 years and it produces India's finest ROBUSTA COFFEE which is used as a blend by International roasters and grinders. One has to learn from the behaviour of this particular bush . It is resistant to almost all insects and pests and responds to very little water without upsetting the quantity as well as quality of coffee. On the other hand the selection Robusta variety 274 is heavily dependent on large amounts of irrigation for flowering.

Age of the Bush

Age is a crucial factor in flower response. In Robusta Plantations , as age increases , the flower bearing and fruit setting also increases. Generally plants above thirty years consistently yield good crops. In young plants , less than six years, the flower buds are nipped because it results in unnecessary stess on the plant.

Moisture

Excessive moisture results in the imbalance of growth regulators and promoters and a particular hormone responsible for vegetative phase comes into play. This drastically reduces the number of flowers.

Flower opening

Under such conditions the bush appears healthy, but the productivity suffers. On the other hand if it rains during the flower opening period, then water gets inside the bud and it starts to balloon up. The flower in such a situation will not set.

Pollination

Honey bees and butterflies are the primary pollinators.

Honeybees pollination

Wind AND MOISTURE also helps to a certain extent. Pollination takes place within five to seven hours after flower opening.

Fertilization

Is completed within 48 hours after pollination.

4

Nursery Practices

Coffee may be grown from seed or from cloned plants in the form of cuttings, grafts or tissue cultured plants. Arabica coffee is most commonly grown from selected seed unless there are special reasons for using clones. A number of steps are necessary for production of good seedlings.

Seed Selection

Arabica coffee should be grown from fresh seed of the recommended varieties. Seed loses viability within three months and should not be used after that period unless properly stored at low temperature and high humidity.

Select ripe healthy fruit from the required variety and from plants that have good productivity, low or no incidence of rust and good cup quality. Pulp cherries, ferment for one night, wash clean, and dry the parchment slowly in shade on raised platforms or trays with good air movement for two to three days. The moisture content of the seeds should not fall below 10%, otherwise the viability will be seriously affected. The seeds should be sorted to eliminate those that are small or abnormally shaped or are infested with pests.

When to Start the Nursery

New seed should be planted as soon as possible after harvest. The longer it is stored, the lower the percentage of germination and the smaller the plants will be at the time of transplanting. If possible, coffee nurseries should be started in December in Lao.

Calculate the Amount of Seed and the Area Required

As coffee seed rapidly looses viability, store the seed in cool moist conditions (such as the bottom of a refrigerator). There are 3000 to 4000 coffee seeds per kilo. The recommended planting density is 3333 plants/ ha at a spacing of 2 x 1.5 m for Lao. To calculate the area for a nursery you need to know:

— the area to be planted;

— plant spacing;

— the number of plants per hectare;

— how many seeds per kg;

— the germination percentage of the seed.

Building Shelter and Beds

Select a frost and flood free area with access to a suitable water supply. Completely fence the area to keep out domestic livestock.

Shade House and Plastic Tunnels

Coffee seed is very slow to germinate in December and January (the coldest months) and clear plastic/ polyethylene should be used to accelerate germination and plant growth. Figure 1 illustrates the stages of coffee seedling development.

Construct a shade house with timber poles and a roof about 1.8 m high. The top of the shade house needs to be covered with either assorted plant material such as bamboo slats or branches, or commercial plastic shade cloth to give about 50% shade.

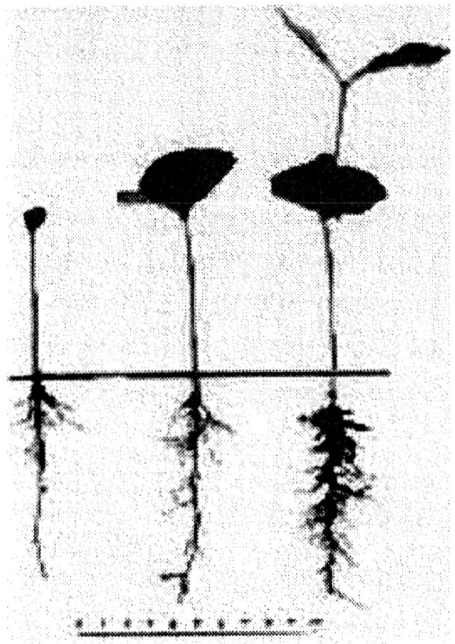

Figure 1. Stages of coffee seedling development

To achieve faster seedling growth during cold weather, plant seed in a clear plastic/polyethylene tunnel beneath the shade. The tunnel is the width of sowing beds and about 75 cm high. Use bamboo hoops for the framework to support the polyethylene sheet cover. The seedbed must be fully and tightly enclosed or temperature inside the tunnel will not increase.

Seedbeds

— Use wooden planks, bricks or bamboo as sides for seed beds which should be about 20 cm high and 1 m wide. Fill beds with a soil and sand mixture of 50% forest soil and 50% river sand. Red soil by itself is too compact for a good seedbed.

— Level the soil to the height of the sides of the seedbed.

Seed Planting

Water the seedbed before planting.

— Using a pointed stick, make furrows 12 mm deep across the bed and 100 mm apart.

— Plant seed flat side down, with seeds 25 mm apart within the row.

— Cover seed with soil mixture - seed should be about 12 mm deep after planting.

— Cover beds with rice straw mulch to give extra heat and to retain soil moisture.

— Water gently. Make sure the seed is not exposed when watering.

As germination time is highly dependent on soil temperature, it may take from 30 to 50 days before shoots appear. Use of plastic/polythene tunnels to retain heat will speed up germination.

Germination

Germination is induced by placing the seeds in a sufficiently moist environment to absorb water. Depending on temperature and moisture, the cotyledon leaves develop after four to six weeks. See figures 2a and 2b for germinating process.

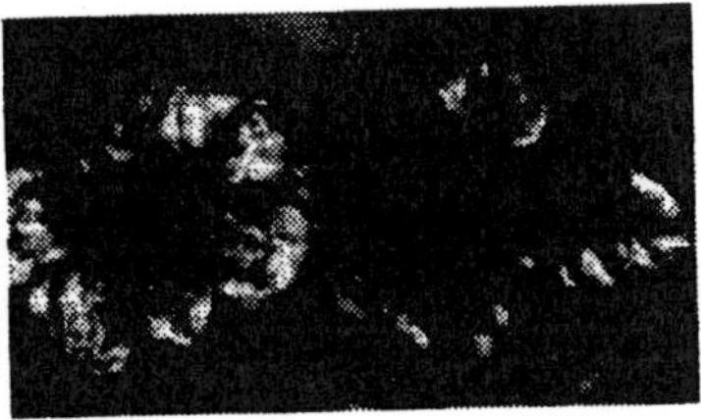

Figure 2. Germinating process

Germination is first seen in the appearance of the radicle three to four weeks after sowing. The hypocotyl (the part between soil and cotyledons appears 20 to 25 days later and carries the seed which is still covered in its parchment, out of the ground. Shortly afterwards, when this light covering is detached, the two cotyledon leaves open. These cotyledon leaves look very different from ordinary leaves - they are oval-shaped with undulating edges and 20 to 50 mm in diameter. At the same time, the terminal bud appears and produces two primary leaves - they are opposite and in pairs. The cotyledons will now die having completed their nutritional role.

The root system develops actively in the first weeks of germination; the taproot penetrates deeply into the soil and forms a great number of roots and rootlets.

The first lateral branch appears four to six weeks after emergence; the plant will then have 5 to 11 pairs of leaves. These branches are opposite in pairs at alternate perpendicular points along the main axis. The primary branches have buds at each node that will develop either into secondary branches or, under certain conditions, into flowers.

Do not let the soil dry out, when seedlings are developing. However, take care and do not over-water as seed can suffer from disease problems such as damping-off. At a height of 200 to 300 mm, the young plants are ready to be transplanted.

Transplanting

Depending on temperature, coffee seedlings are ready to be transplanted from the nursery bed into poly bags about two to three months after sowing. There are four steps in the process.

— Prepare the potting mixture.
— Choose the seedlings.
— Plant seedlings in bags.
— Care for the seedlings.

Prepare Potting Mixture

Strong black plastic/polyethylene bags with drainage holes should be used. Bag size should be at least 100 mm x 250 mm when filled with soil.

A mixture of fertile topsoil and manure or compost can be used. All soil, manure and compost should be sieved. The following mixture could be used:

— 5 x 20 L tins of topsoil.
— 1 x 20 L tin of good quality, dry cattle manure or compost.
— 200 g of rock phosphate or 0:20:0 NPK ratio fertilizer.
— 200 g of dolomite.

Thoroughly mix the ingredients and place in the black plastic bags. This amount will fill about 40 bags.

Selection of Seedlings

— Use the best seedlings with a straight tap root. Discard seedlings with either a bent taproot (J root) or those with few root hairs.
— Do not use larger seedlings (with more leaves than the matchstick stage) as these will be too slow in growing.
— Do not use diseased seedlings.

Plant Seedlings in Bags

— Planting should be done in cool, cloudy weather.

— Thoroughly water the soil-filled bags to settle the soil before planting.

— Lift the seedlings using a stick or trowel to prevent breaking the roots.

— Make a hole about 50 mm deep using either a small stick or a finger (Figure 3).

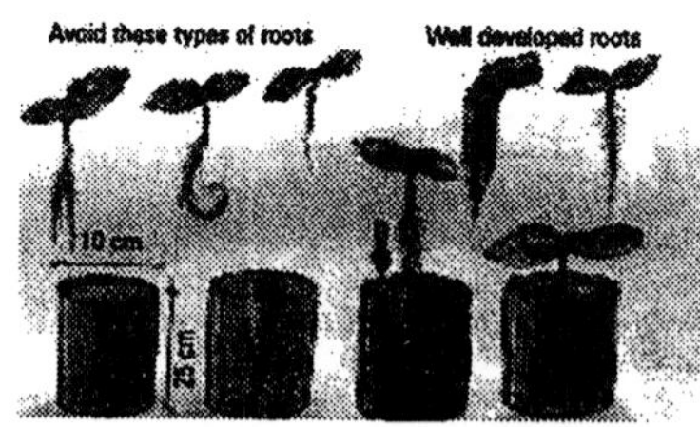

Figure 3. Planting the seedlings into plastic

— Insert seedling in the hole and then lift the seedling slightly to open out the roots.

— When planting, make sure that the taproot is not bent.

— Plant seedlings to the same depth as they were previously planted in the seedbed.

— Water seedlings well.

— Make sure the bags are well supported all around and in between so they do not fall over. Use a bamboo or wooden frame to contain the bags and keep them packed together.

Care for Seedlings

— Remove weeds regularly.

— If soil becomes hard, soften it by using a trowel to

break up big, hard clumps of soil into smaller pieces.

- — Water as required to keep the soil damp. Don't over-water as this can cause damping-off - a disease caused by a fungus that will kill the plants.
- — At three months, apply urea (46:0:0) at 60 g/10 L of water. This is enough for 100 seedlings. Apply every 15 days. If leaves become dark green, stop the procedure.
- — If you do not use a chemical fertiliser, apply a small amount of finely crushed dry manure around the plants.
- — Check seedlings every day to make sure they remain free from pests and disease. Remove bags with diseased, dead or damaged plants.
- — Continue to keep plants in shade. Two months before field planting, gradually remove the shade to sun-harden the plants.
- — As the plants grow, separate the poly-bags so there is sufficient space for the developing plant to spread. If bags are not separated, the plants grow tall and weak.

Diseases and Pests

The two common diseases occurring in the nursery are:

Damping-off that appears as areas of dying plants. Damping-off is caused by a soil-borne fungi often found in old, diseased potting mixture, over-watering, too much shade or not enough space between plants. Damping-off can be avoided by proper preparation in the nursery. It is also important that new soil is always used in the nursery beds. If the disease is found, immediate drenching with either Benlate (Benomyl) or Captan can be carried out.

Always read the label on the chemical pack and follow directions.

Cercospora is a fungus, which develops when plants are under stress caused by too much shade, too much sun, nitrogen deficiency, over-watering or over-crowding. This can be avoided by following good management practices. Immediate control measures involve using copper sprays. Always read the label on the chemical pack and follow directions.

Green coffee scale can also be a problem in the nursery. Scales severely affect plant health as the they suck the sap from the leaves. Keep the area free from ants and spray with spraying oils or Carbaryl or use traditional methods of control.

5

Plant Nutrition

Nutrients are recycled within the environment. A 'closed' environment such as a rainforest, recycles its own nutrients and is more or less self-sufficient. However, where plants are grown in a commercial situation, it is necessary to replenish the nutrients that are removed from the system. Without additional nutrients in some form of fertiliser, coffee yields will remain very low as nutrients are removed with the coffee beans. Unshaded plants of dwarf, high-yielding varieties such as Catimor, will quickly develop dieback and die if adequate nutrients and water are not added to the soil. Plants with mild to moderate dieback will recover with timely good fertilising, watering and weed management.

In India, it was found that for every 6,000 kg of ripe coffee cherry (1 tonne of green bean) removed from the plants, approximately 40 kg nitrogen (N), 2.2 kg phosphorus (P) and 53 kg potassium (K) must be replaced yearly.

There are 16 natural elements (nutrients), that are essential for plant growth (table in the next page). Three elements (carbon, hydrogen and oxygen) make up 94% of the plant tissues and are obtained from air and water. The other 13 elements are obtained from the soil and are divided into two broad categories - 'macro' and 'micro'. These terms do not refer to the importance of the elements; macronutrients are required in greater amounts than micronutrients for normal plant growth.

Essential minerals and their role in the coffee plant

Mineral/ Element	*Chemical symbol*	*Main requirement/use by the plant*
Macronutrients		
Nitrogen	N	Plant growth; proteins; enzymes; hormones; photosynthesis
Sulphur	S	Amino acids and proteins; chlorophyll; disease resistance; seed production
Phosphorus	P	Energy compounds; root development; ripening; flowering
Potassium	K	Fruit quality; water balance; disease resistance
Calcium	Ca	Cell walls; root and leaf development; fruit ripening and quality
Magnesium		Mg Chlorophyll (green colour); seed germination
	Micronutrients	
Copper	Cu	Chlorophyll; protein formation
Zinc Zn		Hormones/enzymes; plant height
Manganese	Mn	Photosynthesis; enzymes
Iron Fe		Photosynthesis
Boron	B	Development/growth of new shoots and roots; flowering, fruit set and development
Chloride	Cl	Photosynthesis; gas exchange; water balance

Soil and Leaf Analysis

To help determine the best nutrition practices, soil and leaf analyses are recommended. While Lao currently does not have access to these services, in nearby Thailand the Mae Jo University in Chiang Mai and Department of Land Development (DLD) can offer fee-for-service analyses. The FAO project used Mae Jo University for the soil and leaf analysis survey of Arabica coffee farms in the Bolovens in 2005.

In order to standardize procedures between farms, years and personnel involved, the following practices are suggested for soil and leaf analysis.

Soil Sampling

— Remove surface litter (leaves, etc.) before sampling. Do not scrape away soil.

— Take samples to a depth of 150 mm with soil auger or spade.

— Place soil in a clean bucket.

— Sample from a minimum of 20 sites across a block of two to four hectares.

— Thoroughly mix each soil sample collected and then sub-sample to reduce volume for sample bags.

— Properly label all samples and laboratory sheets.

— Clean the auger or spade after sampling each of the sites.

— Do not sample after fertilizer application. Scrape away any fertilizer/lime residue from previous applications before taking a sample.

— Do not sample next to shade trees.

— Areas of different tree size, age, soil types, fertilizer or other major differences should be treated as separate samples.

— Samples need to be dried before sending for analysis. If laboratory ovens are unavailable, spread out each sample on a paper bag or plain paper and dry slowly on raised benches under shade and protected from rain. Samples are usually air dry in four to five days.

If possible, soil samples should be taken once per year before flowering.

Leaf Sampling

— Sample the third or fourth pair of leaves from the tip of an actively growing branch. Do not count new leaves if they are not fully expanded.

Figure 1. Leaf sampling

— Sample at the same time/growth stage each year, before flowering.

— Sample a minimum of 40 trees per block across a block size of two to four hectares.

— Sample diagonally across the block.

— Sample average trees only. Do not sample obviously sick, excessively healthy or odd/unusual coffee trees.

— Sample in the morning where possible when leaves are the most turgid (full of water).

— Use clean hands. Do not smoke while sampling and make sure hands are free of fertilizer, soil etc.

— Do not sample when leaves are wet as the paper sample bags will break!

— Do not sample after any application of foliar fertilizer sprays.

— Areas of different tree size, age, soil types, fertilizer or other major differences should be treated as separate samples.

— Properly label all samples and laboratory sheets.

— Samples are to be stored in paper (not plastic) bags. Keep leaves cool but do not freeze!

— Samples need to be dried if they are not sent for analysis within one to two days. This is normally done at the laboratory at 60 to 65°C until dry and brittle.

Pre-flowering is preferred sampling time if only one sample is taken each year. More frequent sampling (every four months) is highly desirable for large plantations, especially if nutritional problems occur. A soil and leaf sampling survey on 15 properties has recently been conducted on the Bolovens Plateaux; results were not available at the time of publication.

Optimum Leaf and Soil Nutrient Levels

Once the soil and leaf samples have been taken, it is important to analyse the results and compare them to levels that have been determined as optimum in coffee plantations around the world in order to devise a nutrition programme for the coffee.

Optimum leaf nutrient levels

Nutrient	*Optimum range*	*Nutrient*	*Optimum range*
N (Nitrogen)	2.5 - 3.0%	Na (Sodium)	< 0.05%
P (Phosphorus)	0.15 - 0.2%	Cu (Copper)	16 - 20 mg/kg
K (Potassium)	2.1 - 2.6%	Zn (Zinc)	15 - 30 mg/kg
S (Sulphur)	0.12 - 0.30%	Mn (Manganese)	50 - 100 mg/kg
Ca (Calcium)	0.75 - 1.5%	Fe (Iron)	70 - 200 mg/kg
Mg (Magnesium)	0.25 - 0.40%	B (Boron)	40 - 100 mg/kg

Optimum soil nutrient levels

Nutrient	*Suggested optimum soil levels*
pH (1:5 soil/water)	5.5 - 6.0
Organic matter (Walkley Black)	1- 3 %
Conductivity (1:5 soil/water)	< 0.2 dsm
Nitrate nitrogen (1:5 aqueous extract)	> 20 mg/kg. Leaf tests more relevant
Phosphate (Colwell or bicarb)	60 - 80 mg/kg
Potassium (Ammonium acetate)	> 0.75 mg/kg
Sulphur (KCl-40)	> 20 mg/kg
Calcium (Ammonium acetate)	3 - 5 meq/100 g
Magnesium (Ammonium acetate)	> 1.6 meq/100 g

Contd....

Contd....

Aluminium (Potassium chloride extract)	Unknown but very low
Sodium (Ammonium acetate)	< 1.0 meq/100 g
Chloride (1:5 aqueous extract)	250 mg/kg
Copper (DPTA)	0.3 - 10 mg/kg
Zinc (DPTA)	2 - 10 mg/kg
Manganese (DPTA)	< 50 mg/kg
Iron (DPTA)	2 - 20 mg/kg
Boron (hot calcium chloride)	0.5 - 1.0 mg/kg (sandy loams)
	1.0 - 2.0 mg/kg(clay loams)
Cation exchange capacity	3 - 5 sandy soil
	> 10 heavy soil types
Cation balance	Potassium (< 10%)
	Calcium (65 - 80%
	Magnesium (15 - 20%
	Sodium (< 5%)
	Aluminium (< 1%)
Calcium: Magnesium ratio	3 - 5

Fertiliser Programme

Coffee soils in Lao PDR are low in a number of essential plant nutrients; therefore these must be supplied to promote high yielding, high quality coffee. Manure, bio-fertiliser, cover crops, compost, legume tree leaves and shoots and chemical fertilisers all supply nutrients.

Manure and compost such as coffee pulp and husks have a low nutrient content. When utilised as a source of nutrients, they must be used in large quantities to supply sufficient nutrients for coffee plants. Manure and compost help improve soil structure and organic matter.

Chemical fertilisers are higher in nutrient content than organic fertilisers and are a more effective method of applying nutrients. For optimal results, it is best to apply a combination of manure and compost and chemical fertilisers. At present, there has been little or no soil and leaf analysis services available for Lao coffee growers. When such services are available, a detailed coffee fertiliser programme can be devised. Meanwhile the following fertiliser programme is suggested for Arabica coffee in Lao.

Year	*Time*	*Application*
Year 1	(Up to 12 months in the field)	Before rains finish
	September	30 g/tree of NPK 15-15-15
Year 2	April/May (with first rains)	30 g/tree of NPK 15-15-15
	July	30 g/tree of NPK 15-15-15
	September	30 g/tree of NPK 15-15-15 500 g/tree of Dolomite
Year 3	April/May (with first rains)	60 g/tree of NPK 15-15-15
	July	60 g/tree of NPK 15-15-15
	September	60 g/tree of NPK 15-15-15
Year 4	April/May (with first rains)	90 g/tree of NPK 15-15-15
	July	90 g/tree of NPK 15-15-15
	September	90 g/tree of NPK 15-15-15
		500 g/tree of Dolomite
Year 5	Onwards	
	April/May (with first rains)	120 g/tree of NPK 15-15-15
	July	120 g/tree of NPK 15-15-15
	September	120 g/tree of NPK 15-15-15

Explanation

1g N = 1,288 g N (Urea)

1g Ca = 1,399 g calcium oxide

= 1,780 g calcium carbonate

1g Mg = 1,658 g magnesium oxide

1g S = 3,750 g magnesium sulphate

Higher yielding coffee plots may require 25% more fertiliser.

Use lime or preferably, dolomite (Ca + Mg) at 500 g per plant every two years and apply before the end of the rainy season. Use the last rains to wash the lime into the soil or water in well by hand or irrigation. The following table shows the nutrient uptake and consumption by different parts of coffee tree (expected yields / ha: 1000 kg green beans).

Nutrient Uptake

	Elements (kg)					
Parts of tree	*N*	*P*	*K*	*Ca*	*Mg*	*S*
Roots	15	2	25	9	2	2
Branches	14	2	20	6	3	1
Leaves	53	11	45	18	7	3
Fruits	30	3	35	3	3	3
Total	112	18	125	36	15	9

It is obvious from this table that leaves need the major part of the uptake - more than the flowers or fruits. However, nutrients are returned to the soil when the leaves drop. The early years of root development are very important as branches and roots store nutrients for a long time.

Nutrients accumulated in the fruits will be removed when cherries are harvested. This loss needs to be compensated by the addition of fertilizers, organic manures, leaf fall or prunings and leaves from shade trees. Recycling of pulp to the soil after composting can help to reduce the additional (chemical) fertiliser needed.

Fertiliser Placement

Spread fertiliser evenly on the soil around the drip line (the outside edge of the canopy) of the coffee tree, as this is where most feeder/hair roots are found. Keep fertiliser at least 100 mm from the stem of the plant; fertiliser applied closer than this can damage the coffee tree.

Manure: The minimum amounts to apply are:

Year 2	0.7 kg/tree
Year 3	1 kg /tree
Year 4	2 kg /tree
Year 5onwards	2.5 kg/tree

Legume shade trees, ground covers and suitable intercrops supply nutrients and organic matter through litter and leaf fall and through prunings added as mulch to the surface of the soil.

Symptoms of Nutrient Deficiency

The overall rate of coffee growth and production depends on the least available plant nutrient. Plants will grow and produce only as much as the least available nutrient will allow them to. It does not matter how much of the other nutrients are available to the plant because it is the least available nutrient that limits growth and development. This is well illustrated in the following 'Barrel Analogy', where the barrel can hold only as much water as the shortest plank will allow (Figure 2). This is known as the 'Law of the Minimum' and is explained thus:

- The level of water in the barrel represents the level of crop yield that is restricted by the most limiting nutrient, nitrogen. When nitrogen is added, the level of crop production is controlled by the next most limiting factor (in this example, potassium).

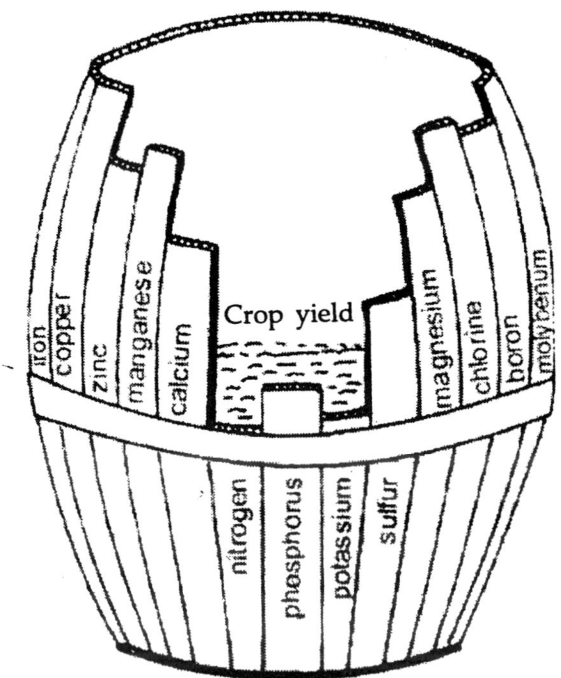

Figure 2. 'Barrel Analogy' using nitrogen as the least available nutrient

Poor nutrition is a major cause of coffee dieback. Plants lacking sufficient N (nitrogen) and K (potassium) suffer from dieback, especially where there is poor shade cover and insufficient water. Low soil calcium and phosphorus will hinder root development and contribute to dieback. Dieback causes loss of yield and when severe, plants can die, especially high yielding, dwarf Catimor varieties.

Each nutrient has unique deficiency symptoms. These are briefly described below.

Coffee nutrient deficiency symptoms

Symptoms originating in older leaves or generally on the whole plant.	*Deficient nutrient*
A. Uniform yellowing over whole tree or light yellowing between the leaf veins.	

Contd....

Contd....

Lower leaves exhibiting slight yellowing, young leaves remaining darker green; faint yellowing between the veins of older leaves at advanced stages; small dead spots may be present.	Phosphorus

B. Localised dead tissue or yellowing between the veins on older leaves.

Initial yellowing on the leaf edges followed by development of dead spots. Dead tissue increases until the whole leaf edge is covered. The veins and midrib remain green.	Potassium
Faint yellowing on leaf edges with sunken, yellow-brown to light brown dead spots developing in a wide band along leaf edges; yellowing between veins evident in affected leaves, particularly along the midrib.	Magnesium
Yellowing in older or middle leaves; mottling, stippling between veins; necrotic spotting along main vein.	Manganese
Bright yellow mottling between veins; leaves wither, curl and margins collapse; leaves distorted and narrow; older leaves affecter first. Rare deficiency.	Molybdenum

Symptoms originating in younger leaves near shoot tips	*Deficient nutrient*
A. *Uniform yellowing over whole leaf or faint yellowing between leaf veins; plants with sparse vegetative growth.*	
Leaves rapidly becoming pale green; new leaves uniformly pale green with a dull green	

Contd....

Contd....

sheen. Entire plant becoming pale green, with sparse vegetative growth; leaves becoming yellow-green at advanced stages; whitish veins may be present in lower leaves.	Nitrogen
Leaves light green to yellow-green, with faint yellowing between veins; deficient leaves retaining shiny lustre.	Sulphur
Whole plant may show symptoms.	
B. *Sharp yellowing between veins of youngest leaves; older leaves*	
Leaves expanding normally, with vein network remaining green and clearly visible against the light green to yellow-green back ground; background becoming nearly creamy white at acute stages.	Iron
Leaves not expanding normally; narrow, often strap-shaped; veins visible against a yellow-green background; failure of inter-node to elongate properly, giving plants a compact appearance.	Zinc
C. *Bronzing, mottling or death of youngest leaves; dieback of terminal buds.*	
Leaves bronzed along edges, cupped down-ward; new leaves dead; eventual dieback of shoot tips.	Calcium
Youngest leaves light green, mottled, with uneven edges and asymmetric shape; new leaves with dead spots or tips.	Boron
Young leaves die back, chlorosis sets in; leaves curl and roll.	

Contd....

Contd....

Shoots are weak and restricted; may be rosetted. Not common if copper sprays are used in nursery and for leaf rust and Cercospora in field.	Copper

6

Pruning Coffee Tree

Pruning

Arabica coffee should be grown as a single stem system. Pruning is required to:

— supply good healthy wood for the next season's crop;

— maintain the correct balance between leaf area and crop;

— prevent overbearing and dieback;

— reduce biennial bearing;

— maintain good tree shape.

Desuckering

Year 1	Desucker to maintain a single stem system and avoid competition from suckers. Remove 'fly crop' fruit (early fruit which compete with strong plant/root development) as they appear.
Year 2	Desucker to remove drooping primary branches that touch the ground. Cut back to nearest secondary branch. Remove secondary branches within 20 cm of the main stem.

Contd....

Contd....	Remove all fruit as they appear (fly crop).
Year 3	Trees should be allowed to crop in the third year.
	Cap the main stem by cutting above a side primary shoot at about 1.6 m from soil level.
	Desucker to remove drooping primary branches touching the ground. Cut back to nearest secondary branch.
	Remove secondary branches within 20 cm of the main stem.
	Maintain a maximum number of well-spaced secondary branches on each primary branch.
	Remove all dead, weak and spindly pest or disease damaged branches.

As plants grow, they can become too crowded and suffer loss of production. Alternative trees can be stumped by cutting off at knee height (0.5 m from soil level). When these trees are producing again after two years, stump the remaining trees.

Rejuvenation Pruning

A regular rejuvenation pruning is needed (normally at six to seven years depending on tree vigour and yield pattern), to maintain a source of new fruiting wood. Unless trees are renewed, yield will decline over the following years.

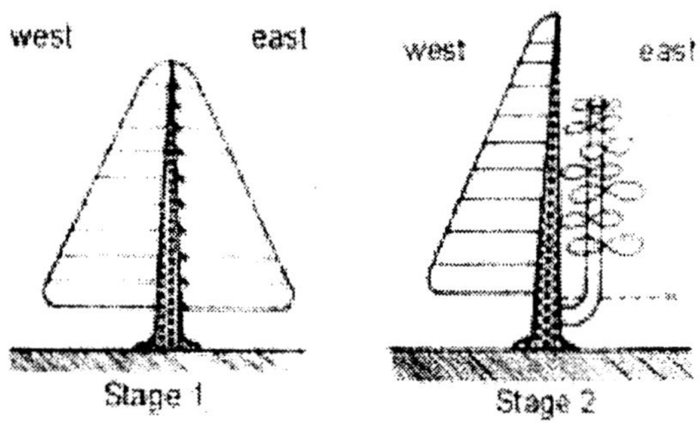

Contd....

Contd . . .

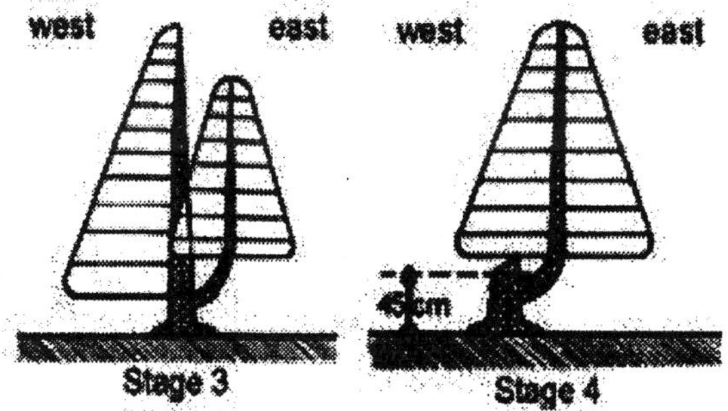

Figure 1. The four stages in side pruning a coffee tree

Two rejuvenation methods are used:

— Side pruning

— Full stumping

Side Pruning

This involves removing one side of the tree, training a new sucker and then removing the other side of tree two years later. This method is recommended for all growers, as only 50% of the crop is lost for the two-year period.

Figure 2. A coffee tree after being side.

Two years before stumping, remove all branches on the eastern side of tree after harvesting. Select a new sucker approximately 300 to 450 mm from the soil level, and train the shoot by thinning as described for a new planting (Stages 1 and 2) until bearing a crop (Stage 3).

Two years later, stump the older stem above the new stem. Cut at a 45° angle - do not cut straight (Stage 4).

Full Stumping

Full stumping involves cutting the tree back to knee height (500 mm from soil level) and developing a new stem from the stump (Figure 3).

Figure 3. Developing a new stem from the stump

This is not recommended, as the crop will be lost for one and most often two years.

Irrigation

Where possible, supplementary irrigation in the dry season will help maintain plant health and maximize yield potential. In Lao, coffee has a water requirement of about 20 to 25 mm per week, which must be supplied from either rain or supplementary irrigation. The amount of water required per hectare for irrigation is about a third to a half less if supplied by drip or under-tree micro-irrigation to the area covered by the plant leaf canopy. Remember that coffee needs to be water-stressed for about four to eight weeks before flowering to give a strong uniform flowering. Do not water trees during this period.

Intercropping

Inter-planting young, non-bearing coffee with vegetables, annual food and cash crops, partly compensates for the high investment cost of coffee establishment, reduces soil temperature, smothers weed growth and supplies the soil with additional nitrogen (legumes) and organic matter when crop residues are turned back into the soil.

Food and cash crops suitable for intercropping include cabbage, peanut, rice, mung bean, vegetables, green beans, maize, upland rice, pigeon peas and pineapple. Keep a distance of 60 cm between the coffee and the intercrop to avoid nutrient and water competition. In some instances with coffee at lower altitudes, pepper vines may be trained up some of the shade trees.

In Bolovens Palteaux, various fruit trees such as durian, guava, lychee and macadamia are sometimes substituted for legume trees.

7

Coffee Industry in India

The coffee industry worldwide is still being haunted by uncertainties though in terms of exports and composite indicator prices there are signs of recovery as per the latest (October 2004) report of the International Coffee Organization (ICO). Total coffee exports globally stood at 7.02 million bags in October 2004 which is the first month of coffee year 2004/05. This signifies an increase of 8.63% over comparable period in 2003 when exports stood at 6.46 million bags.

Total exports during the 12-month period (November 2003 to October 2004) is up by little over 1 per cent to stand at 88.05 million bags compared to 87.11 million bags in the same period in 2003. Exports of Arabica variety conffee in October 2004 amounted to 4.79 million bags signifying an increase of 10.21 percent from the volume recorded in 2003 (4.35 million bags). Robusta exports totalled 2.23 million bags - up by 5.39% over 2.12 million bags, achieved during the comparable period last year.

The ICO composite indicator price recorded over 10 percent increase in composite indicator prices- from 65.19 U.S. cents per lb on 12th November to 72.23 cents on 15th November, 2004. This is the highest level since July 2000.

The Other Milds Indicator Price also set a record on 30 November 2004 with 100.62 US cents per lb. The last time Other milds price breached 100 cents was on 19th July 2000 signifying an increase of 48.4 per cent since the beginning of the year. According to ICO, the average for the decade 1990 to 1999 was 116.08 cents and for the decade 1980 to 1989 139.94 cents per lb.

Volume & Value of Exports

	2000		2001		2002		2003	
	Vol.	Value	Vol.	Value	Vol.	Value	Vol.	Value
Colombian Milds	11.16	1.42	11.67	1.02	11.37	0.98	11.71	1.03
Other Milds	27,06	3.20	22.09	1.83	21.31	1.71	20.45	1.70
Brazilian Naturals	18.32	1.88	22.09	1.42	24.65	1.31	23.75	1.51
Robustas	32.62	1.68	33.50	1.12	30.23	1.12	29.31	1.32
TOTAL	89.17	8.17	90.15	5.39	87.56	5.12	85.22	5.56

At a coffee conference in India in early November (2003) the Executive Director of ICO, Mr. Néstor Osorio observed that the bottom line is that there are only a few measures that could directly address the supply-demand balance. On the supply side – in view of the political and technical difficulties of supply-management schemes – two policies were possible: (a) to create awareness – best achieved through the ICO – in national and international bodies of the danger of embarking on any projects or programmes which would further increase supply; and (b) working to increase the benefits accruing from value-added products rather than traditional bulk commodity exports.

The greatest potential for restoring balance, he pointed out, lay in promoting market development measures to increase demand. These should include: (a) support for the ICO Quality-Improvement Programme as a means of improving consumer appreciation of coffee as well as having an associated effect of removing some sub-standard coffees from the supply side of the world coffee equation; (b) action to increase consumption in coffee-producing countries themselves, which should have a number of positive effects such as providing an alternative market outlet, increasing producer awareness of consumer preferences, stimulation of small and medium enterprises, etc., as well as acting to increase demand; (c) action to enhance knowledge and appreciation of coffee in large emerging markets, such as Russia and China, following the successful ICO campaigns in the 1990s; and protecting consumption levels in traditional markets through quality maintenance, development of niche markets and dissemination of positive information on the health benefits of coffee consumption.

According to latest estimates ICO received from its member-countries indicates to a total coffee production during 2003-04 crop year at 101.55 mn bags compared with 119.59 mn bags in the previous crop year signifying a steep decline of 15.08 percent. Production in 2001-02 crop year stood at 109.27 mn bags. India's coffee production in 2003 is provisionally estimated at 4.50 mn bags -1.25 percent lower than the previous year's estimated 4.56 mn bags. It is to be noted that for five consecutive years, world coffee production remained at above 100 mn bags.

The production during 2003-04 is expected to vary between 100.1 mn bags and 102.4 mn bags. Even USDA estimate which normally speaks on higher side also points to a figure around 106.4 mn bags for 2003-04. Taking into account significant changes in Brazilian coffee production (estimated), the world coffee production in 2003-04 is estimated at 100.1-102.4 mn bags. Various initiatives suggested by the ICO to boost up consumption are yet to impact the market the way they should have. The ICO Executive Director's continuous campaign and appeal to both the importing as well exporting countries seem to have started yielding results.

The consumption in coffee importing countries did improve by 3.07 percent- 63.10 mn bags in 2003 from 61.22 mn bags, while consumption in exporting countries improved by 1.80 percent from 27.15 mn bags in 2002 to 27.64 mn bags in 2003. The export earnings in 2003 are provisionally estimated at US$ 5.56 billion- up US$ 440 mn. However, the opening stocks in exporting countries stood higher in to stand at 20.64 mn bags compared with 19.50 mn bags in 2002. However, stocks of green coffee in importing countries was marginally lower at 20.09 mn bags in 2003 compared with 20.12 mn bags in 2002-2003

The only area of marginal relief is opening seems to be opening stocks in exporting countries which stood at 20.01 mn bags in 2002-03 crop year compared with 21.09 mn bags in 2001-02 crop year and 21.63 mn bags in 2000-01 crop year. It is expected that for sustainable development, the Common Code for the Coffee Community will be finalized by December 2004. This will help develop a global code for the sustainable growing, post-harvest processing and trading of mainstream green coffee.

The Board also considered a report on the establishment of a Sustainable Coffee Partnership, an initiative which complements the Common Code, and which is being developed by the International Institute for Sustainable Development with UNCTAD. Meanwhile the two-day (September 18-19, 2004) second World Coffee conference which is scheduled to be held in Salvador, Brazil will take up all these major issues relating to the problems and prospects of the coffee industry.

India accounts for about 4.5 percent of world coffee production and the industry provides employment to 6 lakh people. Among the coffee growing states, Karnataka accounts for 70 percent of country's total coffee production followed by Kerala (22 percent) and Tamil Nadu (7 percent). Europe accounts for about 70 percent of India's total coffee exports. Of this again, 70 percent is shipped *via* Suez Canal. Major Indian coffee importing countries include Italy, Germany, Russian federation, Spain, Belgium, Slovenia, US, Japan, Greece, Netherlands and France.

In last few years Indian coffee industry is witnessing flat export performance. However, the latest ICO estimate shows that India's exports of coffee in 2003-04 was up 30.94 percent at 4.14 mn bags compared with 3.16 mn bags exported in 2002-03. Special Coffee term loan package announced by the Indian federal government and various other financial booster doses and innovative marketing strategies to a great extent insulated Indian coffee industry from the major shockwaves that swept through the international market severely affecting the bottom line of the companies and resultant impact on the farmers at the grassroot level. India's domestic coffee market is estimated at around 55,000 tonnes to 60,000 tonnes.

Coffee production in India in 2003 is estimated to be 1.25 percent lower than the previous year. However, if 1997 is taken as the benchmark, though in terms of volume India has registered marked growth in exports to 3.40 mn bags in 2002-03 coffee year, export earnings in real terms nose-dived 51 percent at US$ 206.52 million in

2002 from US$ 421.47 mn in 1997. Compared with previous year, exports of coffee during 2002 calendar year was down 15.17 percent - from US$ 243.47 mn to US$ 206.52 mn.

Indian industry is expecting marginal improvement in price realisation in the international market. The quality and aroma of its Arabica variety of Indian coffee has an edge over others in the global coffee market. The Indian Coffee Board is quite sensitive to the plight of the country's coffee industry and has been consistently following possible corrective measures to make the domestic industry feel least impact of adverse international market condition. The restructured Coffee Board reflects the government's frame of mind whereby it would like to go along with all segments of the industry. The inclusion of representatives from Tata Coffee, Hindustan Lever, Nestle India, Barista Coffee and others is a pointer to the changing strategy of the Coffee Board.

Meanwhile, at the international level, to defuse the prolonged crisis, an international high level round table on the coffee crisis, was organised by the ICO and the World Bank where the rich countries have been urged to share the burden of the present crisis that has affected the living standards of 125 million people, mainly in small holdings in developing countries. The USA has been urged to rejoin the ICO. Besides, rich nations have been urged to reduce their agricultural subsidies and tariffs in order to allow potential diversification in those coffee-producing countries willing to move to other crops. ".. we will need to decide whether there is still a role for supply management to address the crisis.

In the light of the universal acknowledgement of an imbalance in the market, we also need to pursue the promotion of consumption through various means, including the improvement of quality, and to reduce dependence through diversification", says Nestor Osorio, ICO Executive Director. "there is no simple solution, there is no silver bullet.

Opening Stocks of Coffee in Exporting Countries
(In million bags)

Crop Year Beginning	Columbian Milds	Other Milds	Brazilian Naturals	Robustas	Total
1990	7.14	5.37	28.89	14.27	55.66
1991	7.67	4.94	28.41	11.95	52.97
1992	8.99	4.95	26.58	14.05	54.57
1993	6.70	3.19	23.65	8.98	42.52
1994	3.66	3.20	25.67	7.81	40.35
1995	6.21	4.21	21.90	7.37	39.69
1996	6.53	3.02	19.14	4.83	33.52
1997	4.41	2.17	17.91	5.29	29.78
1998	4.14	2.42	13.88	5.37	25.80
1999	3.30	2.04	13.00	5.07	23.40
2000	2.59	2.47	11.17	5.41	21.63
2001	1.96	2.61	10.79	5.74	21.09
2002	2.05	3.17	8.54	5.74	19.50
2003	1.90	2.86	9.57	6.31	20.64

India's state wise Coffee Production
(2002-03 Post-monsoon Forecast)

State/District	2002-03 Post-monsoon Forecast		
	Arabica	Robusta	Total
Karnataka			
Chikmagalur	43850	31125	74975
Coorg	23950	68625	92575
Hassan	17250	6500	23750
Sub Total	85050	106250	191300
Kerala			

Contd....

Wyanad	75	54050	54125
Travancore	650	7650	8300
Nelliampathies	450	1550	2000
Sub Total	1175	63250	64425
Tamil Nadu			
Pulneys	6475	275	6750
Nilgiris	1350	2800	4150
Shevroys (Salem)	2900	0	2900
Anamalais (Coimbatore)	1500	450	1950
Sub Total	12225	3525	15750
Non-traditional Areas			
Andhra Pradesh & Orissa	3300	0	3300
North Eastern Region	175	125	300
Sub Total	3475	125	3600
Non-conventional Areas	200	0	200
GRAND TOTAL	102125	173150	275275

There is a consensus that an integrated package which includes improvement in coffee quality, increase in consumption in non-traditional markets, strengthening of the bargaining and marketing powers of producing countries and support to diversification should be implemented", so feels Kevin Cleaver, Director of the Agriculture and Rural Department (ARD), World Bank.

8

Coffee Ecology

Indian coffee plantations are steeped in a tradition of growing coffee under shade by a variety of trees. The coffee plants are surrounded by awesome tree canopy. Hundreds of species of herbs, shrubs, spices ands rare animals share the same habitat. The energy flows; in such a complex system, have distinctive characteristics and needs to be carefully monitored and tailored to suit the requirement of the farm.

Actinomycetes in Coffee Plantation Ecology

Coffee plantations occupy an important part of the Western Ghat landscape. This biological hot spot is home to thousands of species of rare and endangered plant and animal life. Hither to scientists were busy mapping the biotic flora and fauna that was visible to the naked eye. This hot spot also consists of rare and important invisible microorganisms, yet to be discovered by mankind. In fact, scientists have just skimmed the surface. They need to dig much deeper.

Each coffee mountain is unique and distinct and harbors a secret world of microbes. Individuals have different markings to recogonise the species. It is important to understand that shade grown Indian coffee farms provides a natural environment for the growth and proliferation of various microbial groups. For a number of reasons, the microbial inhabitants are in constant touch with both the macro flora and other inhabitants of the coffee mountain. It is this very complexity that has saved the evergreen coffee mountain from changing into a desert.

The micro flora has established a set of rules that favors the build up of a desired species, such that the dynamic state of the entire coffee range is maintained at a level characteristic of the flora. More importantly, the biological equilibrium among and between microorganisms is regulated to a large extent by the overhead canopy of a three tiered shade system which is unique only to India.

Heterogeneous tree populations, not only provide regulated shade to the coffee canopy, but due to the characteristic feature of leaf and fruit shedding at various intervals along with the incorporation of various crop residues favors the build up of both primary and secondary micro flora.

Decomposition of figs inside coffee farms

Many soil microbiologists have reported that the decomposition of a number of natural products is occasionally more rapid in mixed microbial populations rather than with the introduction of pure cultures. The explanation for the phenomenon is obscure but it nevertheless points out to the fact that the coffee mountain provides an excellent environment for the growth and proliferation of billions of microorganisms and they in turn contribute to the well being of the coffee mountain. Microorganisms clearly demonstrate a certain safety in numbers.

Actinomycetes have evoked much curiosity since the discovery of microorganisms by Antony Van Leeuwenhoek, in the 17th century. Actinomycetes are microscopic soil microorganisms and are known to play a very supporting role in the degradation of organic matter in coffee habitats.

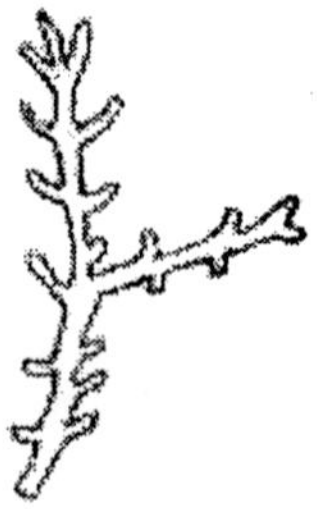

Waksmania

These micro organisms have characteristics common to both bacteria and fungi and yet possess sufficient distinctive features to classify them into a separate category. Actinomycetes produce slender, branched filaments that develop into a mycelium. The filament may be long or short, depending on the species. They form an aerial mycelium, much smaller than that of fungi and many species produce asexual spores called conidia. In fact the leathery or powdery appearance of actinomycetes colonies is due to the production of conidia.

In abundance, they are second only to Bacteria. The resemblance of actinomycetes to bacteria is because the actinomycetes species contain peptidoglycan in their cell walls and possess flagella similar to that of bacterial flagella. In addition actinomycetes are sensitive to antibacterial antibiotics and not antifungal antibiotics. Actinomycetes are also sensitive to lysozyme. Actinomycetes differ from fungi in their cellular composition. They do not possess chitin and cellulose which is found in the cell wall of fungi.

Immediately, after the very first showers inside the coffee mountain, the red earth smells of a musty odor and farmers are confused as to the origin of the odor. They can be rest assured that the odor is the consequence of the presence of actinomycetes.

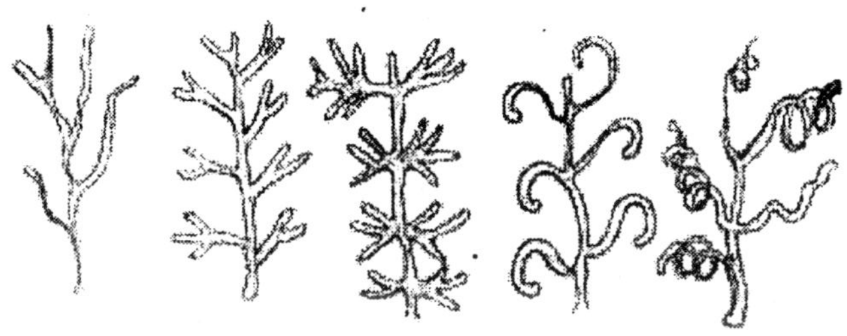

Streptomyces

Many soil scientists have identified the compound or compounds responsible for the earthly odor. The streptomycete metabolite, known as GEOSMIN is largely responsible for the earthly odor. However, other volatile products secreted by streptomyces may also be responsible for the characteristic smell.

Involvement of Actinomycetes:

— Degradation of lignin

— Degradation of organic matter

— Degradation of chitin

— Formation and stabilization of compost piles

— Formation of stable humus

— Production of antibiotics

— Combine with other soil microorganisms in breaking down tough plant and animal residues

Major Group of Actinomycetes:

I. Streptomycetaceae:

— Streptomyces

— Microellobosporia

— Sporichthya

II. Nocardiaceae:

— Nocardia, Pseudonocardia

III Micromonosporaceae:

— Micromonospora

— Microbispora

— Micropolyspora

— Thermomonospora

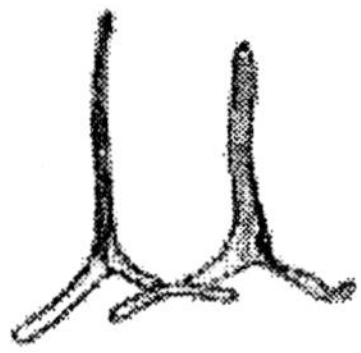

Thermoactino myces

— Thermoactinomyces
— Actinobifida

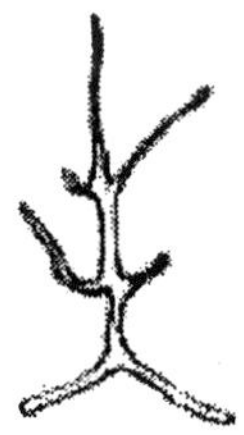

Thermopolysp ora *Streptosporangium*

IV. Actinolanaceae:

— Streptosporangium
— Actinoplanes
— Planobispora
— Dactylosporangium

V. Dermatophilaceae:

— Geodermatophilus

VI. Frankiaceae:

— Frankia

VII. Actinomycetaceae:

— Actinomyces

Growth and Development of Actinomycetes

Microorganisms experience changes in mood depending on a host of conditions.

Organic Matter

The addition of organic matter to coffee soils stimulates

the multiplication and activity of actinomycetes. Actinomycetes belonging to the genera Thermoactinomyces and Streptomyces are commonly observed in compost pits. Pastures or open grasslands converted into coffee plantations have a relatively low population of actinomycetes, primarily due to poor organic matter content. Established farms with abundant tree cover promote the activity of actinomycetes.

Coffee farmers can improve the count of the actinomycete population by simply adding organic amendments like cattle, poultry and sheep manure along with crop residues.

pH

Alkaline and neutral soils are more favorable for the development of actinomycetes. The optimum pH range for the activities of actinomycetes is in the range of 6.5 to 8.0. They cannot survive in acidic Ph. In soils, with Ph less than 5.0 they are almost absent.

Coffee farmers world wide have an important lesson to learn in the control of certain plant diseases, by simply observing the behavior of different actinomycete species to varied ph levels. Alexander Martin, a leading authority on soil microbiology is of the opinion that Ph of less than 5.0 has practical application in the control of certain plant diseases produced by streptomyces; that is, acidification of the soil is used to suppress the pathogen. He further states that even continuous applications of ammonium fertilizers without lime suppresses the actinomycetes, since the ammonium is oxidized to nitric acid by microbial action and the resultant fall in pH leads to unfavorable growth conditions of the pathogen. Liming generally has a beneficial effect because vegetative development is favored by neutral or alkaline conditions, the population being most abundant in soils of Ph 6.5 to 8.0.

Moisture

Actinomycetes are mainly aerobic and as such enjoy well aerated soils. Water logged soils with 80 to 90 % moisture is detrimental for the survival of actinomycetes. If proper drainage is not maintained inside the coffee plantations, especially during the rainy season, then conditions leading to Water logging destroy the population of actinomycetes.

Soil Depth

The percentage of actinomycetes in the total microbial population increases with the depth of soil. However, they are also found in surface soils.

Temperature

The ideal temperature for the growth of actinomycetes is in the range of 25 to 30 degree centigrade. As such most actinomycetes are mesophilic, however, thermophilic actinomycetes play an important role in the transformation of various organic residues inside compost pits. The most common genera of actinomycetes inhabiting the soil are the Streptomyces, Nocardia and Micromonospora.

Chemical Composition

The actinomycetes cell has a carbon content of 45 % and a nitrogen content of 10 %. Lipid content varies from 12 to 65 %. Some species have hexosamine in the cell wall to the extent of 2 to 18 %. The protein and amino acid content of a select few species, namely streptomyces and Nocardia are as follows: Alanine, Methionine, Valine,

Arginine, Lysine, Leucine, hexosamine, Isoleucine, Glutamic acid, Diaminopimelic acid, Threonine, Asparagine.

Actinomycetes in Coffee Environment

The coffee habitat has a profound influence on the qualitative and quantitative actinomycete flora. Especially, in summer, during prolonged droughts, the coffee soils are exposed to high temperatures and subjected to long dry spells. These harsh environmental conditions favor the survival of actinomycetes, because of the production of conidia which can withstand desiccation and high soil temperatures.

Actinomycetes are heterotrophic; hence depend on the availability of organic substrates for their growth and development. Nutritionally, Coffee farmers need to understand the sequences of organic matter decomposition. Initially, it is the bacteria and fungi that are active in attacking the organic substrates.

Actinomycetes develop at a much later stage of plant residue decomposition. Especially, when nutrients are scarce and bacterial and fungal populations are at low ebb, actinomycetes are more prominent. The organic fabric of the coffee mountain is constantly acted upon by various species of actinomycetes.

Many researchers have observed cellulose decomposition by many species of actinomycetes in pure culture, but the rate of decomposition is invariably slow. Many species are known to degrade proteins, lipids, starch, inulin, and chitin, cellulose and hemicellulose. A few strains, belonging to the order Actinomycetales are known to synthesize toxic metabolites.

Potential of Microorganisms

The microorganisms are present almost everywhere and with their short generation time they can be easily tailored to benefit the plantation ecology. They are efficient, invisible, do not occupy precious space and most importantly silently work day in and day out. Microorganisms can literally occupy any inhospitable terrains right from volcanic ash soils to the depths of oceans and the upper reaches of the atmosphere. With satellite technology scientists are probing our planetary system for the presence of microbial life.

Life on earth evolved from a primordial organic soup and microorganisms were the first inhabitants of planet earth. The organic reactor of life was in a soup of water. Complex forms of life arose from this primal ooze. Without their presence and role in breaking down complex organic and inorganic molecules and feeding it into food chains the earth and oceans would be filled with garbage and toxic wastes making it impossible for human habitation.

Isolation of Actinomycetes

Since actinomycete colonies are slow growers compared to bacteria and fungi, they get easily masked in culture plates when grown on ordinary nutrient media. Hence, the easy way out to isolate these microorganisms from soil is by using differential media. Actinomycetes are capable of growing in media containing low nitrogen; hence, media like Ken-Knight's, egg albumin or Conn's medium are used to isolate them.

Plant pathogenic actinomycetes such as Streptomyces scabies are isolated by using tyrosine-casein-nitrate agar medium. In general, actinomycetes are cultured in yeast extract or Czapek's medium.

Role of Fungi in Coffee Plantation

Fungi are a group of diverse and widespread unicellular and multicellular eukaryotic microorganisms. Fungal species are commonly found in soil, in water, on plant debris, and as symbionts, parasites, and pathogens of animals, plants, and other microorganisms. Saprophytic species are important in the decomposition of plant litter and in the recycling of organic matter. Scientists have reported that in nature fungal colonies have been known to continue growing for 400 years or more.

Many fungi are involved in symbiotic associations with various biotic partners within the coffee mountain. Classical examples are mycorrhizae, lichens, and mycetocytes. The systematic study of fungi dates back to the early part of the 17th century. The ancient Greeks and Romans were experts in wine fermentation but they had no modern tools to understand and imply that fungi were largely responsible for these transformations.

Fungus (pl.fungi) constitutes a group of living organisms devoid of chlorophyll. Fungi resemble plants in structure and are capable of utilizing inorganic nitrogen compounds but appear to be more related to animals in requiring oxygen in their metabolism and eliminating carbon dioxide. They have definite cell walls, usually non motile, (they may have motile reproductive cells) and they reproduce by means of spores.

Next to Bacteria, fungi are the most dominant group of microorganisms in soil. The primary role of fungi in coffee plantations is to degrade the vast amounts of complex organic molecules generated from time to time due to the addition of mature plant residues, green manures, organic manures, leaf litter, and convert them into simpler compounds necessary for plant growth and development. Coffee farmers need to understand that most of the biomass available on the coffee farm is relatively free from toxic materials and can be used as stimulants for growth and development of beneficial fungi.

Fungi are important agents in the bio degradation of cellulose, hemi cellulose, starch, pectin, organic acids, disaccharides, fats and lignin which is particularly resistant to bacterial degradation. Fungi are known to adapt themselves to even the most complex of food materials. Fungi are largely responsible in the formation of ammonium and simple nitrogen compounds in soil. Many species of fungi form symbiotic association with plant roots and help in plant growth. Fungi are capable of utilizing simple substances and build them into compounds of higher molecular weight bearing great complexity.

Many fungi are known to produce substances similar to humic substances there by energizing the humic content and organic matter content of the soil. The biomass and organic matter on the forest floor is composed of polysaccharides of varying complexity. Fungi use starch as an excellent source of carbon, chitin is used both as a carbon and nitrogen source by few species of streptomyces. Lipids are attacked by moulds. Complex proteins and polypeptides are further broken down with the help of enzymes.

Fungi consist of filamentous mycelium composed of individual hyphae. The hyphae may be uni or multinucleate and with or without cross walls. The length of fungal mycelium ranges from 50 to 100 meters per gram of surface soil and few microbiologists have recorded values up to 500 meters. Based on filament diameter, specific gravity and mycelial length the weight of fungi ranges from approximately 500 to 6000 kg per hectare of surface soil. The fungal mycelium spreads like a mat and is closely attached to soil particles. These figures indicate the importance of fungi in coffee plantation ecology.

The coffee farm consists of well drained and well aerated soils for the simple reason that coffee cannot tolerate water logged conditions. Luckily, for the coffee farmer, these soil characteristics, together with a thick cover of leaf biomass favors the growth and proliferation of beneficial fungi. Because of their large diameter and underground net work of filaments they contribute significant amounts of total microbial protoplasm, to the coffee soil economy.

Myxomycota

Myxomycetes are also known as Slime Moulds. The characteristic feature of these slime fungi is that they colonize decaying wood, leaf litter, and other organic residues on the floor of the coffee forest. They profusely produce spores known as Sporangia. During favorable conditions the spores germinate and the fungal mycelium spreads rapidly attacking the organic debri.

Eumycota: Mastigomycotina

They are also referred to as algal fungi. This group comprises several moulds, saprophytes, soil forms, parasites and pathogens. Some are known to be parasitic on insects. The asexual spore bearing structure is called SPORANGIUM. They may germinate directly giving rise to germ tubes which form hyphae and mycelium. The fungus is also capable of sexual reproduction. Some species of fungi are known to be plant pathogenic. E.g. Phytophthora infestans.

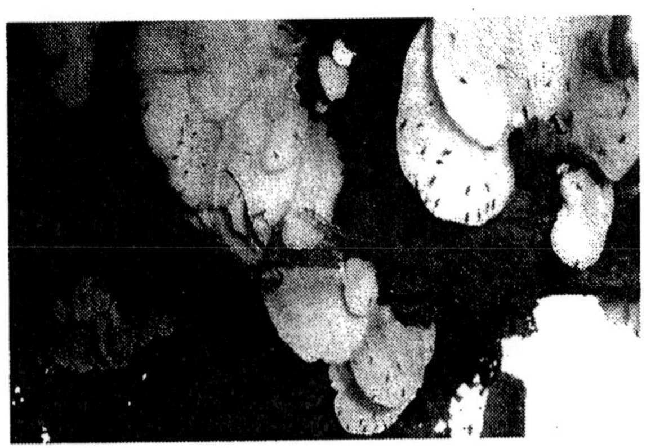

Mastigomycotina

Zygomycotina

Mucor, Rhizopus and Pilobolus are the common genera under this group. They obtain their nutrients from the substrate. In the evolutionary ladder this group consists of advanced morphological characters and sexual forms of reproduction. Zygomycetes and Trichomycetes are the two classes of fungi under this group.

Ascomycotina

They are also referred to as sac fungi because their spores are formed in a sac called ascus. Some fungi are unicellular like yeasts and some like Morchella are highly developed with large fleshy structures. Ascomycetes are very important from the agricultural view point because they are adapted to varied habitats. There are over 35,000 species under this sub division. They are more advanced than the Phycomycetes.

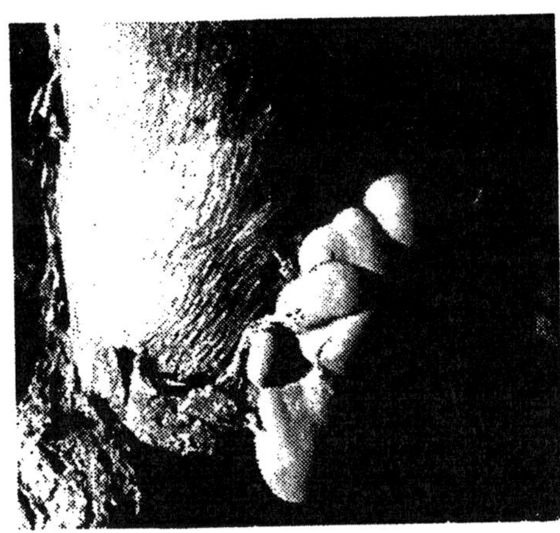

Ascomycotinaa

Yeasts

Yeasts are spherical, ovoid, or rod shaped unicellular fungi. They are widely distributed in coffee soils and sometimes present even on plant parts. They reproduce by the budding mechanism but under certain circumstances a few of them may grow into filamentous forms.

Basidomycotina

This group of fungi too is important from the coffee farmer's perspective because some are saprophytes and others are parasitic. There are more than 15,000 species under this sub division. Basidiomycetes are highly developed fungi. They bear a reproductive structure known as Basidium. The fungi included in this group are mushrooms, toadstools, puffballs, rust, smut fungi, etc.

Deuteromycotina

Fungal species which produce only asexual spores and not sexual spores are considered to be imperfectly understood, hence called Fungi Imperfecti (Deuteromycetes). Some of the fungal species under this subdivision are the Aspergillus, Trichoderma, Rhizoctonia, Penicillium, and Gleosporium. The above mentioned fungi commonly inhabit coffee soils and most of them are of great commercial value in the preparation of microbial inoculants.

Growth and Reproduction

There are two phases in the life cycle of fungi. They may take place simultaneously or in succession. Way back in 1900 Klebs reported that the reproduction in fungi is governed by 4 laws.

Growth and reproduction are life processes, which depend upon different sets of conditions; in the lower organism's external conditions mainly, determine whether growth or reproduction takes place.

Reproduction in the lower organisms does not occur as long as characteristic external conditions which are favorable for reproduction are always more or less unfavorable for growth.

The process of growth and reproduction differ in that growth may take place under a wide range of environmental; conditions than reproduction; growth may take place , therefore under conditions which inhibit reproduction and

Vegetative growth appears to be mostly a preliminary step for reproduction in that it creates a suitable internal environment for it.

Quality of Organic Matter Stauts

The quality and quantity of organic matter plays a major role in the abundance of fungi. The number of filamentous fungi in soil varies depending on the amount of utilizable organic matter present in coffee soils.

Addition of organic matter stimulates fungal activity and is maximum during the initial period of decomposition.

Hydrogen Ion Concentration

Fungi are dominant in acid soils because acidic environment is not favorable for the growth and multiplication of either bacteria or actinomycetes. This special evolutionary mechanism to tolerate highly acidic conditions results in the power of adaptation to hostile environments, including highly acidic soils. However, fungi can also grow in neutral or alkaline soils and some species can tolerate hydrogen ion concentrations beyond pH values of 8.0.

Moisture

The moisture requirement varies from species to species. Soils which have relatively high moisture regimes, suppress the growth and development of certain fungi. Fungi belonging to Basidiomycetes multiply profusely in wood containing less than 20 % moisture. On the other hand the Phycomycetous fungi require very high moisture content for its growth and multiplication. Various moisture regimes determine the fate of spore germination. In general, a high level of moisture facilitates the germination of spores and other reproductive structures of the fungus.

Aeration

Coffee soils which are well aerated contain abundant fungi because most of the fungal species are aerobic.

Temperature

Most fungal species are mesophilic (25 to 35 degree centigrade). However in coffee compost pits, the presence of thermophilic fungi is commonly observed. Thermophilic fungi multiply at 50 to 60 degree centigrade but not at 65degree centigrade.

Season

Since coffee plantations are exposed to different seasons, it has a very deleterious effect on the fungal population. The rainy season provides ideal conditions for the proliferation of fungi but in heavy rainfall regions receiving in excess of 200 inches per annum the growth of fungi are restricted. December and January are winter months in the plantation area and fungi are at low ebb during these months.

Type of Vegetation

Incorporation of crop residues, green manures and carbonaceous materials improves the microbial load, particularly the fungal growth. Certain species dominate initially, but subsequently their numbers decline. Some species maintain high population levels for relatively long periods after the incorporation of plant residues.

Depth

Different species of Fungi are known to occupy different ecological niches. They exhibit selective preferences for various depths of soil. At different soil horizons different species of fungi reside. In coffee soils, fungi are most numerous in the surface layers. A great number of species occur on the surface, and sub surface of coffee soils than in deeper layers of soil. The influence of depth on the distribution and abundance of fungal species may be due to the availability of organic matter and the composition of soil atmosphere.

Farm Practices

Fertilizers, chemicals, weedicides, nematicides, algaecides, pesticides and insecticides influence the type of fungal flora. Ammonium fertilizers are commonly applied in coffee farms. The addition of ammonium fertilizers increases the fungal population and diminishes the bacterial and actinomycete population.

Nutrition

Most fungi are heterotrophs and to date there are no reports of fungi having the pigment chlorophyll to manufacture their own food. The nutrients enter the fungal cell / hypha in solution. Some fungi can use any of the wide range of compounds as sources of carbon and energy while others have highly specific nutritional requirements. A number of fungi require specific growth factors.

Inside the coffee mountain fungi obtain their food either as *Parasites* (infecting living organisms) or by attacking dead organic matter as *Saprobes*. Fungi that live on dead matter and are incapable of infecting living organisms are called *Obligate Saprobes;* those capable of causing disease or of living on dead organic matter, according to circumstances are referred to as *Facultative Saprobes,* and those that cannot live except on living protoplasm are called *Obligate Parasites.*

Coffee Mountain and Chain Reactions

The fungal interactions with the inhabitants of the coffee mountain are essential for the effective transformations of various elements, especially insoluble rock phosphate ultimately leading to release of nutrients for most of the biotic partners. Fungi are also important in the mobilization of nitrogen and phosphorus.

Fungal Spores

Two types of spores are produced by fungi. namely sexual and asextual. Asexual reproduction does not involve the union of sex cells or sex organs. Sexual reproduction is characterized by the union of two nuclei. Asexual reproduction is considered to be more important for the propagation of the species because it occurs several times during a season, where as sexual multiplication generally occurs only once a year.

Most of the fungal species are capable of producing spores in great masses. Because of their light weight, spores are easily dispersed to great distances by wind and air currents. These spores are highly resistant to unfavorable environmental conditions such as high temperatures, heat, extreme cold, desiccation, ultra violet light, extremes of hydrogen ion concentration, and low nutrient supply. The spores are more resistant to heat than the fungal mycelium. The specialized structures that permit survival of the population are conidia, sclerotia, oospores, chlamydospores, sporangiospores, ascospores, sporangia, and rhizomorphs.

The coffee soils are known to contain different species of nematodes. The root lesion nematode, Pratylenchus coffeae is known to cause great economic damage in Arabica plantations. Robusta farms are tolerant. Some species of fungi have developed a mechanism to entrap nematodes and devour them. It has taken millions of years for fungi to develop tentacles helpful in fooling predators. It appears that a normal micro flora has an important protective function against pathogenic and opportunistic microorganisms. Other species of fungi are known to predate over protozoas.

Bacteria in Coffee Plantation

All living organisms are classified as either Prokaryotic (Primitive) or Eukaryotic (Higher forms of life) based on their cellular structure. Bacteria and blue green algae are grouped under prokaryotes and all other organisms are eukaryotes. Bacteria are unicellular microscopic or single celled organisms widely distributed in nature. The greatest benefit in studying bacteria is that it throws light on the evolution of simple cellular systems and the higher forms of life.

The coffee habitat provides a fertile ground generating a host of both macro and micro organisms. The bacteria are the most dominant group of microorganisms in coffee soils. The microscopic analysis of coffee soils shows that there is plenty of room at the bottom for the proliferation of different types of microorganisms because of the rich humus and organic matter content.

Among the different groups, bacteria are capable of harvesting atmospheric nitrogen, solubilisation of rock phosphate and in the transformations of various substrates resulting in periodic supply of available nutrients for plant growth and development. However, coffee farmers need to understand that the types of bacteria and their numbers are governed by the soil type and cultivation practices like addition of chemical manures, pesticides, poisons, type of tillage etc.

In turn the activity of bacteria is influenced by the availability of nutrients, both in organic and inorganic forms. Their numbers are very high in coffee soils with large number of trees compared to open meadows. This is due to the shading nature as well as greater root density and the abundant availability of soil organic matter. Due to the high organic matter content of coffee soils, bacteria decompose the organic matter and in the process acquire energy.

Bacterial cells are so very small that when you think of these fastidious and ubiquitous microbes, you need to think small. They are measured in microns and the equivalent of one micron is: one by thousandth of a millimeter. Hence one needs a fairly high powered microscope to observe these minute wonders. However, they have a remarkable advantage because they have their strength in numbers.

The population of bacterial cells in soils is always great. Due to their rapid growth and short generation time they can quickly act on various organic materials. In harsh environments lacking oxygen the bacteria alone are responsible for almost all the biological and chemical changes. Because of their very small size bacteria have a very high ratio of surface area to volume. Also, since bacteria are single celled microorganisms they absorb their nutrients through their cell membrane, there by exhibiting very high metabolic rates.

The earliest inhabitants of Planet Earth have undoubtedly been the microorganisms. From primitive prokaryotic unicellular microorganisms, evolved the higher forms of life. Microorganisms have thus been the earliest participants in shaping various life processes. Microorganisms have been largely responsible in changing the primordial atmosphere resulting in the formation of gaseous oxygen needed for plant growth.

Fifteen million years of evolution has shaped the coffee forest and today their future is in our hands. Fundamentally, it has been an evolution of skills. The evolutionary ladder points out to the pivotal role played by microorganisms in adapting to harsh environmental conditions, formation of tripartite bonds, break down of complex polysaccharides into simpler molecules and in the process providing the energy needs of the biotic community. Among the different microorganisms, Bacteria are known to play a vital role in the distribution and supply of energy needs of the entire coffee mountain. Decomposition of almost all insoluble salts is mediated by one or the other group of bacterial communities.

Functions of Bacteria

Bacteria are widely distributed along the length and breadth of the coffee mountain. In short they are found almost every where. Bacteria are single celled organisms and in spite of their simplicity are highly efficient. Their numbers decline with depth of soil. A majority of the coffee farmers are unaware that the great majority of bacteria are beneficial and absolutely necessary to convert farm wastes, organic debris and other by products into energy rich compounds needed for plant growth and development.

Plants and animals depend on the fertility status of the soil and this in turn is dependent on the activity of soil microorganisms. Plants cannot directly utilize organic compounds such as fatty acids, lipids, carbohydrates and proteins. Microorganisms are a vital link in the mineralization of organic constituents and provide nutrients in the available form for plant growth and development.

Bacterial cells can withstand long periods of drought due to the protective cover around the cell wall known as *Capsule*. The capsule is a slimy or a gelatinous material and encloses either one cell or a group of cells. At times the bacteria make use of the polysaccharides present in the capsule as a source of reserve food material. The capsule enables the bacteria to avoid predation by larger soil microbes and infection from viral strains. In addition to protection, capsules also play an important role in the attachment of bacterial cells to plant or rock surfaces and in the formation of biofilms.

Bacteria are morphologically grouped into three types. Cell structure is a key element in the characterization of bacteria.

Cylindrical or Rod shaped commonly referred to as Bacilli. They are the most numerous. Bacillus species are known to overcome extreme weather conditions by the formation of endospores that function as part of the normal life cycle of the bacterium. These endospores are resistant to long periods of drought and desiccation. With the on set of favorable conditions the spore germinates and a new bacterial cell grows.

Communication between all life is essential in unfolding the various patterns of life. Without the ability to communicate, life, including the simplest single celled organism, could not exist. Bacteria have the capacity to analyze vibrations in the surroundings and accordingly react. Besides shape and size certain rod shaped bacteria have thin hair like appendages on the outer cell wall known as flagella, which can sense the external environment and constantly send out chemical signals to reach out to other communities. Flagella are believed to be organs of locomotion.

The number and place of attachment vary. When a single flagellum is attached to one end of the rod it is known as *Monotrichous*. If the flagella are attached singly at both ends it is known as *Amphitrichous;* If more than one flagellum or a bunch of them are attached to either one end or both the ends it is known as *Lophotrichous* and if the flagella are covered all over the cell it is known as *Peritrichous*.

Spherical or ellipsoidal bacteria are called cocci. Ellipsoidal bacteria occur in pairs and are referred to as *Streptococci,* when in four cells, arranged in a square they are known as *Tetrads;* when in irregular clusters like a bunch of grapes they are called *Staphylococci* and when arranged in a cubical form known as Sarcinae.

Winogradsky a leading soil microbiologist placed Soil bacteria into two broad divisions.

Autochthonous species

These refer to the indigenous or native species. The population of these bacteria is always uniform and constant in coffee soils because their nutrition is dependent on the native soil organic matter. They multiply rapidly in the presence of large quantities of biomass, organic matter, humus, and other soil amendments having a low C:N ratio. They are pretty tough and resistant to varied agro climatic conditions. They participate in all biochemical functions of the community. The presence of these bacteria is fairly high and their numbers are constant. The presence or absence of specific nutrients does not change their numbers significantly.

Allochthonous species or Zymogenous Bacteria or Fermentive

These bacteria are active fomenters and need nutrients which are quickly exhausted. They are involved in a process in which organic matter is rapidly attacked in successive stages and made available to the plants. At each stage of decomposition a specific group of organism is involved. The bacterial numbers increase rapidly whenever furnished with the special nutrients (leaf litter, biomass, compost) to which they are adapted. On exhaustion of these nutrients their numbers decrease and return with the addition of nutrients. Hence, this group of bacteria requires an external source of energy for their multiplication and growth.

Bacteria in this group include the nitrogen fixers, phosphorus solubilisers, nitrifiers, cellulose hydrolyzing bacteria, sulphur oxidizers, spore forming bacillus and non spore forming pseudomonas.

Environmental Factors

Bacterial numbers, their density, type and composition is governed by the environmental Stimulus. The important factors are listed below.

Aeration

Bacteria are further divided as

- — *Aerobes*: Require the presence of oxygen for growth and metabolic activity.
- — *Anaerobes*: Bacteria which grow in the absence of oxygen.
- — *Facultative Anaerobes*: Develop either in the presence or absence of oxygen.
- — *Aerotolerant Anaerobes*: These bacteria grow under both aerobic and anaerobic conditions.

Moisture

Aerobic bacteria are the main stay in coffee soils and the optimum level of moisture content for their activities is in the range of 50 to 75% of the soil's moisture holding capacity. Coffee soils are inherently shaded by tree canopies as well as by the coffee bush. Hence they remain shaded most of the time. Also, a host of factors result in the availability of moisture throughout the year. For e.g. The south west and the north east monsoon together keep the soil moist for eight months of the year and the remainder months, due to soil conservation practices adopted by the Indian coffee farmer , the moisture is always available for bacterial growth and development.

Water makes up a major component of the microbial cell. Hence it is a key component for the functioning of the cell. The most common problem encountered in coffee soils is not the lack of moisture but the availability of excess moisture which is detrimental for the growth and multiplication of bacteria. Excess moisture limits the supply of gaseous oxygen resulting in an anaerobic environment. Water logging brings about a decrease in the abundance of bacteria.

Temperature

Bacteria are highly sensitive to temperature fluctuations. Apart from growth and development, temperature plays a vital role in the biochemical processes carried out by the bacterial cell.

— *Mesophiles* are the ones which grow well in the temperature range of 25 to 35 degree centigrade. These constitute the bulk of the coffee soil bacteria. For most part of the year the temperature profile in the coffee mountain falls in this range. Some scientists have further divided mesophiles as:

  — *Oikophilic;* Organisms whose optimum temperature is around 20 degree centigrade.

  — *Somatophilic;* Organisms whose optimum temperature is about 37 degree centigrade.

  — *Psychrophyles* are bacteria that love cold and grow at temperatures below 20 degree centigrade.

— *Thermophiles* are temperature loving bacteria and grow best in the temperature range of 45 to 65 degree centigrade. These bacteria are active in compost pits.

Organic matter

The population of bacteria is directly related to the organic matter content of the soil. Due to periodic leaf shedding and availability of huge quantities of carbonaceous materials on the floor of the coffee forest, the bacterial numbers is the largest. Also the coffee farmers incorporate green manures, compost and biomass from time to time which act as stimulants for the growth and proliferation of bacteria.

Acidity

The optimum pH for the growth of bacteria is Neutral EUTRAL pH. Coffee farmers need to keep the hydrogen ion concentration of their soils close to neutral because in highly alkaline or highly acidic conditions the growth and multiplication of bacteria is inhibited. In general in heavy rainfall areas receiving 100 inches and more it is advisable to apply lime or dolomite once every two years and in moderate rainfall regions, once every four years. This practice will not only increase the bacterial numbers but will also enable the coffee bush to take up inorganic nutrients in a more efficient way.

Inorganic nutrients

Application of fertilizers and chemicals greatly affects the bacterial population. Coffee farmer's world wide use ammonium fertilizers as the bulk of fertilizer application. Coffee farmers do not realize that ammonium fertilizers tend to lower the soil pH resulting in acidity due to the microbial oxidation of ammonium to nitric acid. More than the effect of fertilizer, it is the acidity which suppresses the bacterial population. This problem can be easily overcome by split applications spread out over a two week period. More importantly, the application of fertilizer should be carried out when the soil moisture is optimum. It is a proven fact that small amounts of inorganic fertilizers supply the needs of the bacterial community in the form of inorganic nutrients.

Farm Practices

Farm practices also exert direct and indirect biological effects on the coffee farm. Periodic soil disturbance will affect the bacterial population. Addition of organic manures from time to time and incorporating legumes into the soil with proper carbon nitrogen ratio accelerates the build up of beneficial micro flora. However, if soil hardens up over a period of time, then it will have an adverse effect on the bacterial numbers.

9

Soil Water Conservation in Coffee Plantations

In India, coffee plantation regions are characterised by heavy rainfall. However, the amount of rainfall received on any given day is unpredictable inspite of the advances made in weather forecasting. Productivity depends on two critical components, water and soil which vary from one zone to another. Improved soil water conservation inside the Plantation can also lead to greater fertiliser efficiency use.

Studies on the effect of erosion have shown that the downfall of many flourishing empires was primarily caused by soil degradation. If soil loss is greater, than soil development, due to accelerated soil erosion then the agriculture systems in the world will be seriously threatened. The survival of the coffee bush depends on the possibility of providing adequate soil moisture during the dry months. Thus the important question that comes to mind is finding out ways and means of maximising the efficiency of soil water retention for greater productivity and sustainability.

To understand the complex mechanism of conserving precious water, one needs to understand the basics of soil profile. Soil profile plays an important part in the retention of surface water, subsurface water and ground water. Depending upon the physical forces at play and the mode of retention, soil water may be divided into three categories, namely Hygroscopic water, Capillary water, and Gravitational water.

Hygroscopic Water

Hygroscopic water is that water which is bound to the soil clay particles and hence incapable of any movement. It is held so tenaciously by soil particles that plants cannot make use of it. The force with which it is held on the clay particle is reckoned to be 10000 atmospheres. However the osmotic pressure of the root sap of most plants is about 20 to 25 atmospheres and hence plant roots are unable to draw upon this water from the soil.

Capillary Water

Capillary water is water which adheres to the surface of the soil particles as liquid water and is in the form of a thin film. As moisture conditions increase the thin films of water form rings around the soil particles and is available to the coffee plants. The molecules of capillary water are free and mobile and are present in the liquid state. Even though capillary water is retained on the soil particles by surface forces, it is not held firmly as hygroscopic water. Yet it is held so strongly that gravity cannot separate it from soil particles.

Gravitational Water

Due to heavy rains the amount of water in the soil increases beyond its maximum capillary capacity, and this additional water appears as free water in the soil pore spaces. This water is beyond the force exerted by soil particles and therefore comes under the influence of gravity. Hence this water is called Gravitational water.

The downward movement of gravitational water through the soil is called Percolation. A part of this water moves through cracks formed in the soil due to shrinkage, or through channels and cavities formed by organic matter decomposition. This water is called *Seepage water*. Many a times due to the pervious nature of the soil, the gravitational water percolates down into the lower layers and drains away. If for some reason the soil layer is impervious, then the percolation water that collects over it is forms the *Ground water table.*

Looking back at the rainfall pattern in most coffee growing regions, the Monsoon is spread over a period of six months. However, July, August, September and October months are known for torrential rains where maximum soil erosion and runoff takes place. This transport of precious soil to river beds is catastrophic. Soil erosion is a destabilising factor in all agro ecosystems.

The amount of water that runs over the surface of the ground is considerable. Depending on the contour of the land, and ground cover the annual average loss varies from anywhere from 15 to 35 per cent of the total rainfall. This is commonly known as surface runoff. This surface runoff it is worth investigating, as it brings about a depletion of soil fertility.

It's more a rule in Indian Plantations where every block has cradle pits (3feetby 1feet by 6 inches width). These cradle pits are for young plantations and in older plantations the pits are larger in size, preferably called trenches varying anywhere between 24 feet long 11/2feet wide and 2feet deep. These pits are dug across the slopes and thereby in spite of any eventful downpour or cloud burst, all the soil and nutrients along with water is retained in these pits.

Generally coffee is grown on red and lateritic soil and red sandy loam and within a short period, all the water inside these pits gets drained down to the lower regions of the soil, finally reaching ground water in the purified form, retaining precious soil and nutrients at the subsurface level which accumulates over the years and is recycled inside the Plantation.. The values of these water pits are two fold. Firstly they conserve water, but most importantly during the formation of these water pits the plantation ensures a host of benefits.

- As the pits are dug the surface feeders of the coffee plants get cut and this gives new vigour to the plant.
- Prevents soil erosion.
- The top soil is spread on to the root surface of the plant and enhances productivity.
- Better aeration for root growth and development.
- Excellent conditioning for better infiltration of water.
- Prevents spread of root borne diseases.
- Pits accumulate organic matter over a period of time which is recycled to the estate.
- Acts as a catalyst in the proliferation of beneficial soil microorganisms.
- Purification of water from harmful pathogens.
- Avoids silting of lakes.

In addition to these trenches and pits, different cultivation practices such as scuffling and cover digging of soil just before the on set of the monsoons, helps in conserving soil moisture. Coffee forests with their dense mulch and organic matter act as blotting papers in absorbing rain into the fragile earth. They indirectly, purify more than 70 per cent of the water that we drink, from harmful bacteria. Even if there is a cloud burst inside the coffee plantation, one can hardly seen any runoff because of the blotting paper effect due to a mosaic of insitu activities within the coffee ecosystem. Due to the sponge effect the humus rich soils in one hectare of forest can store about 6 to 7 lakh tons of water after filtering. The top soil itself can hold 1.2 lakh tons of water. This speaks volumes about the importance of coffee forests not only as precursors in conserving water but also as vital links for the survival of Coffee plantations which in turn act as huge watersheds.

The crucial test to understand the health of the coffee plantations is to avoid cracking up of the soil, which leads to rise in temperature within the microhabitat and there by soil looses its power to sustain the Plantation. Soil microorganisms which give life to the soil are the first to be affected and then the macro organisms which work in tandem with the soil. A degree rise in soil temperature during the summer months due to inadequate levels of soil moisture will result in the death and destruction of millions of beneficial microorganisms. We have observed over the years that inspite of the thick mulch inside the Plantation , if the annual rainfall is much below average, then the soil starts cracking up and this can simply result in an immediate loss of soil vitality and will take years to bring it to proper health.

Generally, the coffee Plantation comprises of a three tier shade system with the top most canopy anywhere between 100 to 150 feet. It is seen to be believed that when there is a heavy downpour the kinetic and potential energy of water can play havoc inside sensitive areas where ground cover is bare minimum. The first event that unfolds is the hammering of the rain drop from a great height directly into the ground and as the continuous pounding of rain drop into the same area , the soil quickly looses its binding strength and slowly is broken down into smaller and smaller particles resulting in the formation of sand. This sandy loam soil does cause enormous harm to the sustaining capacity of the ecologically sensitive and biologically active coffee farm.

Equally, important is soil water conservation in relation to runoff and erosion. Soil erosion is almost universal. After all soil is a finite resource. There is a conscious need to efficiently manage and conserve earth's water resources in a manner that will allow maximum productivity on a sustainable basis. Far from the truth, inspite of advances in modern agriculture, improper soil management and lack of conservation practices have resulted in loss of precious top soil. It is an established fact that soil water conservation and soil erosion are two faces of the same coin. The second most important fact is that major causes for soil erosion is due to unplanned farming activities.

According to Central Soil & Water Conservation Research Training Institute, Dehradun 5334 million tons of soil is being eroded annually. Of this 29 % is being permanently lost to sea, 10% is deposited in reservoirs as silt, 61% displaced from one location to another. This study warns that present annual acreage loss of top soil is approximately 16 tons per hectare; far above the permissible limit of 4 tons per hectare. Future generation of farmers is bound to be affected by this huge loss.

There is good news round the bend for all coffee farmers. Firstly, due to good soil management and cultural practices inside the farm, the accumulation of humus and mulch accelerates and this in turn keeps the soil temperature at an optimum level for enhanced biological activity. We have excavated six big water tanks inside *Kirehully Estate* with an average depth of 18 feet. This observation points to an amazing discovery that within the Plantation zone , hidden underneath the subsurface of the soil are special rocks which grow in a very systematic fashion, either vertically downwards or simply in a horizontal manner and these rocks; identified as *Water Harvesting Rocks.*

The nature of these rocks is that they are striated, pink in appearance and when one holds them to the cheek; it is as if a wet towel is held. There is irrefutable evidence that whenever a pond is dug, if one comes across these rocks, then the success rate of getting water is very high. The beauty of these natural water harvesters is that they keep the biological clock clicking in an orderly fashion by providing sufficient moisture at different depths for trees, insects and microbes. They also provide the much needed aeration at the root zone.

Most Planters have not realised that the Rhizosphere Region (Root Zone) of coffee plants, constitute one of the best moisture reservoirs inside the plantation. Water that infiltrates into this reservoir can be stored with relatively little loss for very long periods because the evaporation losses are bare minimum.

10

Pests and Diseases

The word pest covers mites, aphids, caterpillars, slugs, snails, beetles, flies, moths, butterflies, nematodes, bugs, rats, monkeys, grazers, and humans. There are many pests, these pests have been responsible for wiping out whole plantations.

Insect Pests

Green Coffee Scale

Green coffee scale *(Coccus viridis)* is a common and serious problem. Scales suck the plant sap resulting in reduced growth and crop yield. Sooty mould (a black, loose, sooty-like cover) often develops on leaves. It grows on the sweet exudate from the scales (honeydew) that also attracts ants.

Symptoms

Green oval shaped scales about 2 to 3 mm long. Often found concentrated on leaf veins and tips of new shoots. Infestations then produce spots of honeydew, which become covered with a black sooty mould. Defoliation of badly affected trees can occur.

Control

Preventative

There are a number of natural predators of coffee scale such as wasps, ladybugs and Verticillium fungus. In many instances, these will reduce the level of scale infestation.

Chemical

Mineral spraying oils at 200 ml/ 20 L water applied as a spray to affected plants. Only spray if 10 or more leaves are infested with one or more scales. The spray must completely wet and cover the scales. Do not use automotive oil! Carbaryl 85 % wettable powder at 20 g/ 10 L water applied as a spray. Apply weekly until scales disappear.

Traditional

1 kg strong tobacco per 2 L water. Soak for 2 nights. Then remove tobacco. Add 500 g of washing powder and make up to 20 L. Spray weekly until scales disappear.

Aphids

Aphids *(Toxoptera aurantii)* can occur in large numbers on new shoots in the rainy season. Aphids suck sap from young shoots and cause damage to these developing shoots.

Symptoms

Large numbers of small black aphids (2 to 3 mm long) concentrated on new growth. Often associated with black sooty mould.

Control

Generally not warranted.

Chemical

Neem oil 10 to 20 ml/L, plus soft, finely grated laundry soap at about 7 g/L water.

Stemborers

There are two species of stemborer present in Lao PDR.

Red stemborer (*Zeuzera coffeae*). The adult has white and black spotted wings. The red coloured larvae tunnel through the coffee branches, normally in the upper part of the coffee trees. Branches and the top part of the main stem easily break off, but the tree usually survives.

White stemborer (*Xylotrechus quadripes*). The adult is a black and white banded beetle (about 1 to 2 cm long); the head of the male beetle has distinctive raised black ridges. Adults are active during daylight. Damage is caused by the white larvae, which hatch from eggs deposited in cracks and crevices and under loose scaly bark of the main stem and thick primary branches, especially on plants exposed to sunlight. Young larvae feed on the corky tissue just under the bark, which splits making the stem appear ridged. Later, larvae enter the heartwood and tunnel in all directions, even into the roots.

Symptoms

Wilting of leaves and dead trees or branches. Affected branches are easily broken off. When trees are first infested there maybe evidence of frass (sawdust-like residues) on the ground. The trunk may be ringbarked. The lifecycle of both pests is completed during the rainy season, but often damage is more evident during the dry season.

Larvae remain inside the tree and are normally not seen. Usually damage is not economically important, although individual trees can be lost.

Control

Preventative

- — Less damage occurs under conditions of good shade.
- — Higher altitude (above 800 m.a.s.l.) seems to reduce the incidence of infestation.
- — Burn affected trees or branches with borers inside.
- — Do not plant trees with twisted taproots. These deformed roots result in weak trees that have been shown to have a high incidence of stemborer infestation.

Chemical

No effective chemical control known. Biological control is not known at this time.

Coffee Berry Borer

Coffee berry borer *(Hypothenemus hampei)* is a relativity new, but very serious problem in Lao. It is causing significant damage, with perhaps as high as 50% yield loss. The adult is a small black beetle (about 2.5 mm long) and covered in thick hairs. The female beetle bores into berries through the navel region. Cherries are attacked in various stages but tunnelling and laying of about 15 eggs occurs only in hard beans. The eggs hatch in about 10 days and the larvae feed on the beans making small tunnels. Beetles in the cherries either on the plant or on the ground, can survive for more than five months.

Symptoms

Fruit drop of young, green cherries. A small hole is evident on the cherry. Cherries that do not drop often have defective, damaged beans.

Control

Orchard hygiene (keeping the area clean, removing dropped cherries, removing carry-over fruit from coffee bushes are suggested), but it is reported to have limited impact and can be expensive. Cherries on the ground and old berries remaining on the trees are sources of new infection.

There are few natural enemies of the borer. One wasp *(Phymastichus coffea)* has shown promise in Columbia, but its effectiveness and that of other wasps is not yet fully known. The wasp may make a contribution in an IPM system. Lao should procure this and other effective parasitoids from Cenicafe in Colombia and technical biocontrol support.

Interest is now focused on the commonly found fungus, *Beauverai bassiana*. Research in South America has shown promising results, but it is not a cheap alternative to chemicals and has to be re-applied. Research is required to develop the best means of bio-control.

Chemical control is difficult as the borer spends most of its life cycle deep inside the coffee berry. Endosulfan 35 EC at a rate of 6 ml/4.5 L of water applied at early fruit set (2 mm cherry size) and later 120 to 150 days after fruit set if required. Cypermetrin and Deltametrin, pyrethroids (0.01%) at 26 ml/15L of water are an alternative, or Chlorpyrifos used at recommended rate on label.

Quarantine: The pest cannot migrate any distance on its own. Do not allow cherries or coffee bags from other farms onto the farm property. Crop bags should be fumigated before being transported to other coffee growing areas. Ethyl alcohol and methyl alcohol at a rate of 1:1 is effective in trapping CBB and can be used most effectively at processing/ washing places to prevent reinfestation. Place traps in the first five rows of coffee growing near the processing area.

Coating pieces of plastic with axle grease and engine oil and attaching these to pulpers and machines in the coffee processing area can also be used to capture CBB. Careful drying of coffee cherry or parchment reduces reproduction of the pest as they cannot survive in coffee beans that are properly dried to 12% moisture.

Mealybug

Mealybugs *(Planococcus* spp.) are small sucking insects (about 3 mm long) covered with a white mealy wax that feed on young shoots and young roots. There are several species similar in appearance to the naked eye. They are generally more of a problem in the dry season when water is lacking. However, serious intestations of mealybug are often found where there has been use of insecticide sprays, especially highly toxic organophosphate sprays. These kill almost all insects, including natural enemies of mealybug.

Symptoms

White waxy colonies are usually found on the underside of tender leaves and in soft stem areas around berries. Also, they are found on young roots near the main root, especially where soil is loose around the trunk. Mealybugs are often associated with a heavy infestation of sooty mould.

Control

Biological

Normally sufficient. In other countries, the most important predator is the mealybug ladybird *Cryptolaemus montrouzieri.* The adults are reddish brown with black wings and about 4 mm long. A parasitic wasp, *Leptmastix dactylopii,* is also very effective. Lacewings such as *Oligochrysa lutea* are also predators of mealybug.

Chemical

Spray Chlorpyrifos on the soil around the tree to kill ants. Ants disrupt the natural enemies of the mealybug. Malathion and Carbaryl sprays can also be effective. Apply according to label recommendations.

Leaf Miner

Leaf miner *(Leucoptera coffeina)* is often present, especially in shaded coffee.

Symptoms

Transparent areas in the leaf; larvae are present on the underside of the coffee leaf. Fully-grown larvae are about 6 mm long.

Control

Normally a minor problem with no control warranted.

Termites

Termites *(Macrotermes* spp.) can be a problem on older coffee and shade trees with dead wood where termites breed.

Control

Plant coffee in clean ground where all tree parts, including roots have been removed. Termites cannot survive as there is no dead wood on which to feed.

Effective pruning of dead wood on coffee trees. Remove all dead wood from the coffee plantation. Permetrin 60 to 80 g/L sprayed on the ground and on base of coffee trees after planting will assist.

Diseases

A number of diseases can affect coffee plants in the nursery as seedlings, in the field while young and later as bearing trees.

Nursery Diseases

Coffee seedlings are susceptible to two main diseases in the nursery - Damping-off and Cercospora leaf spot (brown eye spot).

Damping-off

This disease occurs on young coffee seedlings in the germination bed, after germination and before transplanting. It is caused by a *Pythium* spp. fungus.

Symptoms

Patches of coffee die quickly. Coffee stem is soft and rotten.

Causes

— Soil borne fungi.

— Soil too wet.

— Too much shade (insufficient drying of soil).

— High planting density (too many plants in a small area).

Control

Preventative

— Don't use old soil from nursery beds or bags as disease is soil borne and can be carried over.

— Use new soil for nursery beds and potting-up.

— Avoid over-watering.

— Do not plant seed too close; seeds should be 25 mm apart in rows 100 mm apart.

Chemical control

Soil drenches of either Benlate (Benomyl) or Captan (Follow label directions as formulations differ).

Cercospora leaf spot is a fungus that occurs on leaves when plants are under stress. The fungus can develop both in seedbeds and after plants have been transplanted into bags. It is the most common nursery disease and a sign of poor management.

Symptoms

— Brown spots on leaves gradually expanding with reddish brown margin.

— Spots on both sides of the leaf.

— When there are many spots, leaves appear to have been burnt.

Causes

— Soil too wet.

— Too much shade or too much sun.

— Lack of air movement.

— Lack of nitrogen and potassium.

Control

Preventative

— Avoid over-watering.

— Maintain 50% shade cover.

— Space plant bags to allow air movement.

— Proper fertiliser application.

Chemical

Copper sprays such as the following will give control:

- Copper Cupravit (85% WP)80 g/20 L water
- Copper oxychloride 80 g/20 L water
- Copper hydroxide 40 g/20 L water

Field Diseases and Disorders

There are several field diseases and disorders affecting leaves and berries. Diseases include Cercospora leaf spot (all ages of coffee); coffee leaf rust (all ages but more on bearing coffee); black sooty mould (all ages) and Anthracnose (more prevalent on bearing coffee). The severe disorder, overbearing dieback, occurs on bearing coffee.

Cercospora (Berry Blotch & Brown Eye Spot)

This occurs on the leaf but can also occur on berries where it is known as berry blotch.

Symptoms

- Brown spots on leaves gradually expanding with reddish brown margin.
- Spots on both sides of the leaf.
- Brown sunken lesion on green berries surrounded by a bright red ring (berry blotch).

Causes

- Low leaf nitrogen and potassium. Insufficient shade.
- Stress from drought, sun exposure, poor fertiliser management, excessive weed competition.

Control

Preventative

Maintain well-fertilised plants with 50% shade cover.

Chemical

— Should not be needed with good management.
— Copper sprays such as the following will give control in severe cases on isolated plants:

Copper Cupravit	(85% WP)
	80 g/20 L water
Copper oxychloride	80 g/20 L water
Copper hydroxide	40 g/20 L water

Coffee leaf rust

Coffee leaf rust *(Hemileia vatatrix)* occurs on leaves and can cause leaf drop if severe.

Symptoms

The first symptom is the formation of pale yellow spots up to 3 mm in diameter on the underside of the leaves. As the spots expand, they become powdery and yellow to orange in colour and may reach 20 mm in diameter. Occasionally the whole leaf becomes covered with rust spots.

Older rust spores become brown at the centre surrounded by powdery orange spots. Leaf drop occurs, which if severe, can lead to dieback and berry loss and a loss of both yield and quality. Berries tend to be very small, not fully ripe and turn black.

Causes

Variety: Catimor is rust resistant. Java, Typica and many

other Arabicas are susceptible under poorly shaded conditions and at altitudes of less than 1000 m.a.s.l.

Plant health: Healthy plants are less susceptible.

Control

Preventive

— Continued coffee leaf rust

— Plant Catimor selections or other more tolerant varieties such as good selections of S 795.

— Follow the recommended nutrition programme.

— Plant pure Arabica at high elevation only and always use good shade.

Chemical

Monthly copper sprays (May to October). See label directions for rates.

Sooty Mould

Sooty mould (*Capnodium* spp.) develops when the plant is infested with scale, mealybugs, aphids or other sucking insects.

Symptoms

— Leaves covered with black, powdery soot.

— The fungus grows on honeydew produced by green coffee scale and sucking insects. Ants care for the scales and spread the sooty mould.

Control

Preventative

Reduce levels of coffee scale, aphids and mealybugs by using recommended control procedures.

Chemical

Not needed if sucking insects are controlled. Control the insects, not the disease.

Anthracnose

Anthracnose (*Colletotrichum gloeosporioides* Penz.) is a minor flower, twig and cherry disease. It can cause three different coffee diseases - twig dieback, brown blight of ripening cherries and leaf necrosis.

Symptoms

- Twig dieback - yellowing and blight of affected leaves.
- Twigs wilt, defoliate and die at the tips.
- Brown blight - brown sunken lesions on fully developed cherries which turn black and hard (can be confused with Cercospora).
- Leaf necrosis - round brown necrotic spots up to 25 mm diameter. Worse on sun-burnt or injured leaves.

Control

- Maintain healthy coffee plants.
- Other control measures are not warranted.

Overbearing or Dieback

Overbearing: Plant cannot support the extremely heavy crop. Not a true disease but a physiological problem.

Symptoms

- Severe leaf loss and branch dieback.
- Root dieback.
- Cherries ripen prematurely and become hard and black.

— Dieback causes alternating bearing (heavy crop one year and poor crop the next).

— Plants decline and eventually die if the problem is not corrected in early stages.

Note

— Coffee needs one leaf pair to support five to six berries through to maturity.

— If there are too many cherries and not enough leaves, all the food goes from the leaf to the developing cherry. Leaves then drop off, causing dieback. Some varieties, especially dwarf Catimors, are more susceptible to this condition. Loss of leaf depletes plant carbohydrate reserves resulting in weakened plants.

— Roots also die back, then the tree cannot take up enough nutrients and water, thus more leaves are lost and cherry quality is reduced.

— Plant health decline continues and if plants are not well cared for with adequate watering and nutrients, the plants will succumb and die.

Causes

— Insufficient nutrition.

— Insufficient shade.

— Insufficient irrigation.

Variety

Dwarf Catimors are much more susceptible.

Control

Preventative:

— Once the problem exists it is very hard to break the cycle if it is left too long.

— Maintain good plant health. Maintain good shade (50%). Plant only recommended varieties.

— Use a well-balanced fertiliser programme and apply adequate nitrogen and potassium as recommended earlier.

11

Fertilisers for Coffee Plantations

Use of Bio-fertilisers

Many present-day coffee plantations use chemical fertilisers, especially nitrogenous and phosphatic ones. These types of fertilisers are bad for two main reasons: they are very expensive, and these chemical fertilisers often make use of non-renewable energy resources like fossil fuels, which can deplete nature's precious resources. Furthermore, these synthetic fertilisers can harm Mother Earth due to water pollution. Thus, these chemical fertilisers are disastrous for the fragile ecology of coffee-growing regions.

Many generations of coffee farmers have ignored and abused the soil. The soil imbalance process takes time and the changes in each generation are minute, so no one cared—and many people still don't care. Now, many farmers need to enrich the soil, so they turn to chemical fertilisers.

Faced with a problem of such a enormous magnitude, one can find a easy, yet effective solution for fertiliser needs of Coffee Plantations by just looking at the soil as a major natural resource. Most people think of soil as a dead, inert material. However, from an agricultural standpoint a healthy soil is the lifeline of any nation. Soil itself is a living system with millions of beneficial microbes, acting as factories that provide biological nitrogen and other nutrients to the plant. Consider that 83.3% of the earth's atmosphere is made up of inert nitrogen gas. Microorganisims can convert this atmospheric nitrogen and make it available to the plant in the soluble form (such as ammonia) that the plant can absorb and utilise. This process enriches the soil; thereby enriching the ecosystem.

Microorganisms in the Soil

Microorganisims are present almost everywhere. Microbes are ubiquitous and at the same time promiscuous. The tip of a needle can hold more than 100,000 bacteria; so when you talk about bacteria, you've got to think on a grand scale. The soil acts as a reservoir for millions of microorganisms, of which more than 85% are beneficial for plant life. Thus, the soil is a resilient ecosystem. Good soil consists of 93% mineral and 7% bio-organic substances. The bio-organic parts are 85% humus, 10% roots, and 5% edaphon.

Humus is a product of the synthetic and decomposing activities of the microflora; it exists in the dynamic state. It is under continual attack, yet it is constantly reformed by the subterranean inhabitants. Similarly, edaphon is a world of life and consists of microbes, fungi, bacteria, earthworms, micro-fauna, and macro-fauna as follows:

— 40% fungi/algae

— 40% bacteria/actinomycetes

— 12% Earthworms

— 5% Macrofauna

— 3% micro/mesofauna

Thus, soil microorganisims provide precious life to soil systems catering to plant growth. These microorganisims work *incognito* to maintain the ecological balance by active participation in carbon, nitrogen, sulphur and phosphorous cycles in nature. Soil microorganisims play a pivotal role both in the evolution of agriculturally useful soil conditions and in stimulating plant growth.

For quite sometime the Coffee Planting Community world wide has preferred synthetic fertilisers. They analyse the soil for Ph, electrical conductivity, macro and micronutrients. However, none may have analysed their soils for microbial count. Times are changing, and the role of bio-fertilisers for augumenting the fertiliser needs of Plantation crops is gaining significance. This method is not only safe but preserves the soil for future generations by enhancing and maintaining soil fertility.

Bio-fertilisers in Coffee-Growing Areas

Tropical areas where coffee is grown should use bio-fertilisers for several reasons.

— In the trophics, the restricted availability of major nutrients like nitrogen and phosphorus limits plant growth and yield. Meanwhile super-phosphate fertiliser is expensive and in short supply, but bio-fertilisers can bridge the gap.

— Also, in India for example, phosphate-deficient soils can be enriched with phosphate-fixing soils, which are often widespread in other areas.

— The soil temperature is high in tropical areas, which acts as a catalyst in enhancing microbial activity; thereby increasing the flow of nutrients to the plant. So bio-fertilisers are an efficient fertilising option.

— Bio-fertilisers do not pollute the environment. They are ecofriendly and harmless. Bio-fertilisers address the core issue of supplementing nutrients, without affecting environment.

— Biofertilisers are a low-cost technology for coffee plantations, the majority of which are owned by small-time farmers. A low-cost solution that enriches the soil gives a thrust to economic development without disturbing ecological balance.

Most fertilisers add nitrogen to the soil. This can be done via chemical fertilisers, or through a process called biological nitrogen fixation (BNF). On a worldwide basis it is estimated that about 175 million tons of nitrogen per year is added to soil through biological nitrogen fixation (BNF). The term bio means *living*; so bio-fertilisers refer to living, microbial inoculants that are added to the soil. These bio-fertilisers are products consisting of selected and beneficial microrganisims, which are known to improve plant growth through supply of plant nutrients.

The soil microorganisims used in biofertilisers are: Phosphate Solubilising microbes, Mycorrhizae, Azospirillum, Azotobacter, Rhizobium, Sesbania, Blue Green Algae, and Azolla. Let's go through these groups in a little detail in order to understand their role in bio-fertilisers, which can be used to make rich, living soil that is suitable for coffee plants.

Phosphate Solubilising Microbes

Phosphorus is an important nutrient for plants. There are several microorganisims which can solubilise the cheaper sources of phosphorus, such as rock phosphate. Bacteria like Pseudomonas striata, and Bacillus megaterium are also important phosphorus solubilising soil microorganisims. Many fungi like Aspergillus and

Penicillium are potential solubilisers of bound phosphates. They solubilise the bound phosphorus and make it available to the plant, resulting in improved growth and yield of crops.

Soil phosphates are rendered available to plants by soil microorganisims through secretion of organic acids. Therefore, phosphate dissolving soil microorganisims play some part in correcting phosphorus deficiency in plantation soils. They may also release soluble inorganic phosphate into soil through decomposition of phosphate rich organic compounds. These microbial inoculants can substitute almost 20 to 25% of the phosphorus requirement of plants.

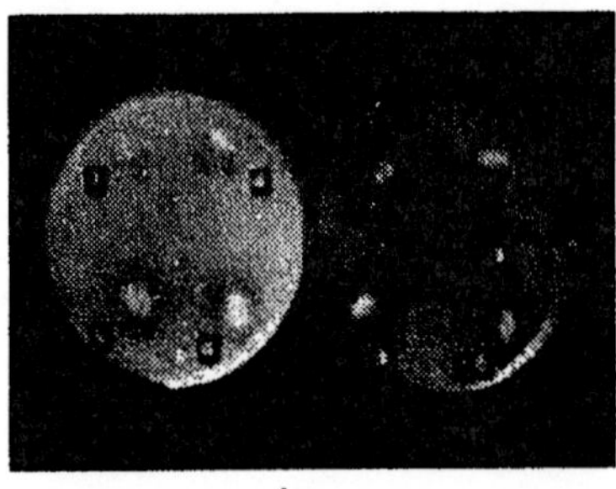

Rock phosphate solubilising microorganisms

Phosphate solubilising microbes can also be inoculated to coffee husk along with rock phosphate while preparing compost to enrich the compost with available phosphorus.

Mycorrhizae

The term "mycorrhizae" refers to fungus associated with plant roots. These fertilisers are divided into ectotrophic and endotrophic or the vesicular arbuscular mycorrhiza (VAM) categories. Most plants depend on their mycorrhizal association for adequate uptake of nutrients (especially the immobile ions such as phosphate, zinc and micronutrients) and survival in natural ecosystems. Mycorrhixal association stimulates branching of the root and increases the absorption surface of the root. Other benefits include tolerance to drought, high soil temperature, soil toxins, and extreme Ph levels, as well as protection against root pathogens. This is why, When trees are introduced to new regions, inoculation of soil with mycorrhizal fungi is a necessary prerequisite for the establishment of the trees.

Azopirillum

Azospirillum are nitrogen-fixing bacteria that lives in a symbiotic relationship in the root cortex of several tropical crops. They stimulate plant growth through nitrogen fixation and production of growth subustances like auxins, gibberellins and cytokinin. It is estimated that almost 10 to 15% of the required nitrogen can be met by Azospirillum biofertiliser.

Azobacter

Azotobacter are free-living, nitrogen-fixing bacteria and are known to produce several plant growth promoting subustances. In addition to nitrogen fixation by these bacteria, they are also known to protect plants against pathogenic microorganisims either by discouraging their growth or by destroying them. These inoculants need more attention in view of their triple action of nitrogen fixation, bio-control, and production of plant growth regulators.

Rhizobium

Rhizobium bacteria, basically form root nodules in leguminous plants and fix atmospheric nitrogen in a symbiotic association. The Rhizobium bacteria gives nitrogen to the plant and the plant gives protection to the bacteria from oxygen damage by harbouring it inside the root nodule.

Sesbania

Many legumes are grown and then turned into the soil while they are still green to enrich soil nitrogen. Sesbania is a green manure plant which forms both root and stem nodules in association with rhizobium and thereby fixes more atmospheric nitrogen. These legumes produce ten times more nodules than other legumes and have a very high capacity to fix atmospheric nitrogen. Apart from enrichment of soil nitrogen, green manuring enriches the phosphorus, calcium, sulphur and other micronutrient of the soil.

Blue Green Algae

Blue Green Algae (BGA) or Cyanobacteria have the ability to carry out both photosynthesis as well as nitrogen fixation. They belong to the order Nostocales and Stigonematales. Algal flakes are grown and then broadcasted.

Azolla

Azolla is a floating fern which harbours a blue green algae in its leaf cavities.

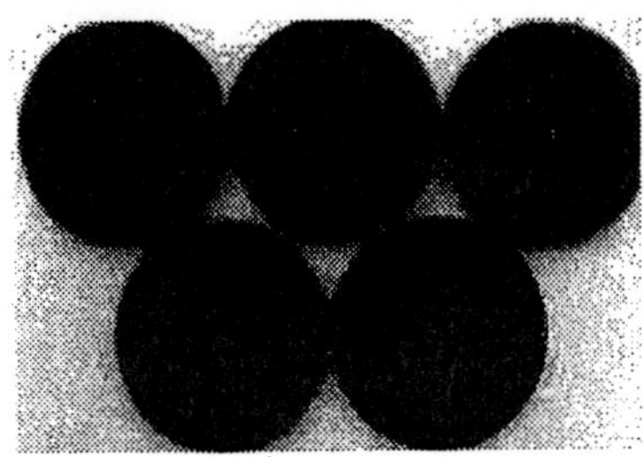

Nitrogen-fixing Azolla strains

The fern multiplies very fast with the symbiotic association of the algae and this rapid multiplication creates a huge amount of biomass on the surface of the water. It is then harvested, dryed and used as biofertiliser.

Neem Based Fertiliser

The British pioneers opened up coffee plantations in specific geographical belts known for their soil fertility and moderate climate. Care was taken to only plant traditional varieties. Organic cultivation of coffee was the norm and a main stay in the farmer's package of practice. Coffee farmers were in a position to harvest sustainable yields without high pest and disease incidence. The present, situation is very different. The farmer's life has become stressful and the coffee farm is burdened with toxic levels of synthetic fertilisers. This perhaps alone will change the secure future of forest grown coffee.

The importance of fertiliser, particularly nitrogen, in increasing coffee yields and quality was proved beyond doubt by researchers and this lead to the widespread and indiscriminate use of nitrogenous fertilisers in boosting coffee yields. Today, Coffee farmers have expanded their area of coffee cultivation together with multiple crops and rely on synthetic nitrogen fertilisers for better crop establishment and productivity.

Urea—Nitregen Fertiliser

Urea is the principal nitrogen fertiliser used by a majority of the coffee farmers. The efficiency with which coffee utilises nitrogen is notoriously low. World wide research clearly indicates that under ideal laboratory conditions, for every dollar of urea (nitrogen fertiliser) applied, under field application, the fertiliser use efficiency is not more than 30 %.

Studies demonstrate that insufficient fertiliser level or inefficient management of applied fertiliser, particularly improper timing of nitrogenous and potash fertilisers, accounts for at least a 500 kg per hectare gap between farmers coffee yields and their potential farm yields. Another important aspect overlooked by coffee farmers : Application of a small amount of nitrogen will bring about low productivity of coffee, where as an excessive amount may decrease yield. Hence, it is very important to use a balanced level of fertiliser at the right time.

Split Dozes of Fertiliser Application

Enlightened coffee farmers are aware of the significant losses of fertiliser nitrogen during soil application. They have devised a simple method to minimise these losses and also increase nitrogen efficiency by adopting the split application of fertilisers. The recommended doze of fertilisers is applied in 4 split dozes. However, this practice has two drawbacks, namely unavailability of labor at the desired time and the expensive labor cost which adds to the cost of cultivation.

Coffee Terrain

Synthetic nitrogenous fertilisers are subjected to dynamic transformations within the confines of the coffee mountain due to varied slopes, different soil, environmental, cropping and management factors. The ecodynamic coffee cube is quite a complicated cube in which many chemical, biological and physical processes affect nitrogen reactions. Nitrogen transformations are also highly complex in nature resulting in significant losses of applied fertiliser through denitrification and ammonia volatilization.

Unfortunately, the causes for these significant nitrogen losses are beyond the control of coffee farmers because of the unpredictability of the rains. Many a times soon after application of fertilisers there is a sudden downpour or an extended period of drought, associated with rise in daytime temperatures.

The potential nitrogen loss mechanisms operating inside the coffee landscape are:

— Ammonia volatilisation (loss 30 to 70 %)

— Leaching (loss 30 to 60 %)

— Denitrification (loss 30 to 60 %)

— Surface run off (loss 15 to 20 %)

— Nitrogen immobilisation (loss 10 to 40 %)

— Ammonium fixation. (Loss 5 to 10 %)

The extent to which each process operates is not clearly understood. The above mentioned losses can be significantly reduced by the application of neem. Neem controls the rate of release of nutrients from nitrogenous fertilisers and also retards nitrification in soil by controlling the activity of nitrifying bacteria.

The role of neem in controlling the release of synthetic fertilisers has been understood for quite some time. Losses of nitrogen from urea and low recoveries from coffee soils can be significantly overcome with the help of neem coated urea, which acts as a slow release nitrogen fertiliser with the objective of making the amount of nitrogen released coincide with that required by coffee. Neem coated fertilisers get mineralised at a slower rate than just urea and also provides residual nitrogen for future growth and development. Also, neem coated fertilisers have a carry over effect which is beneficial for root and shoot development. Neem also plays a significant role in controlling ammonia volatilisation and leaching losses of nitrogen.

More importantly, neem increases the efficiency of nitrogen utilisation by not only coffee, but multiple crops like pepper, orange, and areca nut. Neem also acts as an excellent nitrification inhibitor, there by increasing the efficiency of fertiliser nitrogen. The actual chemical compounds responsible for nitrification-inhibitory properties in neem kernels have yet to be identified. Neem cake treated urea contributes to a remarkable reduction in urea hydrolysis, there by conserving nitrogen. In this system more attention is focused on conserving energy.

Neem consists of proteins. Proteins undergo hydrolysis with the liberation of amino acids which through biological oxidation yield nitrogen in the form of ammonia or elemental nitrogen. Numerous other compounds arise as a result of microbial degradation. Ultimately, there is a simplification of numerous compounds and the synthesis of others. The complex biological structure of neem undergoes a break down into simpler compounds resulting in microbial succession favoring heterotrophic nitrogen fixers. These slow and gradual changes result in the formation of a nitrogen pool which is available for plant growth and development.

Japan and the U.S.A have carried out pioneering work in the use of controlled release fertilisers which are sparingly soluble in water. Ureaform, crotonilidine diurea, sulphur coated urea and isobutyledene diurea are some of the controlled release fertilisers. The prohibitive cost of these fertilisers restricts its reach beyond the use of the common farmer. An inexpensive substitute to the above mentioned controlled release fertilisers is the blending of neem with urea. Neem seeds contain certain lipid associates which act as nitrification inhibitors and when mixed with synthetic urea, increases its efficiency.

Table 1. Controlled release fertilisers

Fertiliser	% N
Crotonilidine diurea (CDU)(Urea+acetaldehyde)	32% N
Urea-form (Urea+Formaldehyde)	38%N
Guanyl urea (GU)	37% N
Sulphur coated urea (SCU)	40% N
Oxamide $H_2NCO—CONH_2$	31.8%N
Isobutylidine diurea (IBDU) (Urea+Isobutylaldehyde) (CH_3) 2——CH=CH—(NH—CO—NH_2) 2	32.2%N
N-lignin (Ammonified lignin)	18% N

Nitrification Inhibitors

Nitrogen losses can also be significantly stemmed with the help of nitrification inhibitors like pyrimidines, acetanilides, isothiocyanates, pyridines and anilines. Two major chemicals which have been commercially produced by the Dow Chemical Company. U.S.A. and the Toyo Koatsu company of Japan are N-SERVE and AM. Unfortunately these chemicals are very expensive and are not available in India.

Synthetic Nitrogen on Soil Microflora

Over two decades reveals that the highly organised systems of living things are super sensitive to chemicals in the external and internal environment. Excessive application of fertiliser results in profound, often drastic effects on soil micro flora and fauna. The beneficial microbes are destroyed resulting in the proliferation of microbes which take up fertiliser nitrogen as an energy source, for cell growth and development. The nitrogen which was supposed to be taken up by the plant is instead taken up by soil microbes resulting in the further loss of nitrogen through immobilisation. Neem coated nitrogen fertilisers stimulate beneficial microbes and leave behind a significant positive effect on crop growth and yield.

Impact of Neem on Synthetic Fertilisers

— Reduces Leaching loss

— Slow release of nutrients from complex fertilisers.

— Improves efficiency of other nitrogenous fertilisers.

— Compatible with a wide range of synthetic fertilisers and chemicals.

— Reduction in urea hydrolysis.

— Inhibits nitrification

— Reduces ammonia volatilisation

— Increases efficiency of nitrogen utilisation by coffee plants. Increase in nitrogen assimilation

— Supplies nitrogen to coffee and other allied crops for extended periods of time.

— Regulates supply of nutrients coinciding with the crop growth and development.

— Conserves higher ammonium nitrogen in soils.

— Drastic reduction in the population of nitrifying bacteria.

— Reduces the population of ammonium oxidisers (Nitrosomonas sp) without affecting nitrite oxidisers (Nitrobacter sp).

Efficiency of Nitrogen in Sprinkling

Robusta coffee farms are invariably subjected to artificial blossom showers during the summer months of February and March during which the daytime temperatures are high. World wide research suggests that at the time of flowering, the plant requires a tremendous amount of energy and the best way of providing this is by applying nitrogenous fertilisers like urea which is readily taken up by the plant.

Uniform berry ripening-Neem coated fertiliser effect

Application of urea during blossom showers increases the flower setting and the subsequent fruit set. However, more than 70 % of the applied nitrogen is lost by volatilisation due to high daytime temperatures. Generally speaking the problems associated with urea include hygroscopicity, rapid dissolution, and ready decomposition to ammonia associated with a temporary increase in soil pH. Also, immediately after sprinkling, due to high moisture status, there is every possibility of urea leaching out beyond the root zone.

For over 25 years we have practiced applying neems mixed with urea and other nitrogenous fertilisers and have successfully arrested the significant nitrogen losses.

12

Monitoring Coffee Plantations

World wide, almost all coffee producing Nations grow coffee on the steep slopes of mountains. This is especially true of shade grown Indian coffee. These coffee mountains are often exposed to high velocity winds and also torrential down pours which result in the leaching of precious nutrients.

Depending on the average rainfall, in a specific agro climatic zone, the hydrogen ion concentration of the coffee soil (pH)varies and needs to be corrected every two to three years for better uptake and assimilation of nutrients. Soil pH is important because of the many effects it has on biological and chemical activity of the soil, which affects the metabolism of your plants.

Importance of pH

pH: It is important to understand the true meaning of exactly what pH is. pH is a true measure of how acidic or alkaline your soil is. Just like we use lbs., kilos to measure weight, we use the pH scale to classify the soil pH. This pH scale ranges from 0 to 14. The lower the number the more acidic, the higher the number the more basic with a pH of 7 being Neutral.

The pH value reflects the relative number of hydrogen ions (H+) in the soil solution. The more hydrogen ions present, compared to the hydroxyl ions (OH-), the more acidic the solution will be and the lower the pH value. If you will notice the hydroxyl ion and the hydrogen ion combined will give you (H_2O) water. So pure water has a neutral pH of 7. Once things are added this balance of ions will shift one way or the other to make the water acidic or basic.

The scale is logarithmic meaning that a soil with pH of 5.0 is 10 times more acidic than a soil of 6.0 and a 100 times more acidic than a soil of 7.0.

pH Meter

This hand-held device is wonderful tool for serious coffee planters who would like to constantly monitor the health of their life supporting soil system by checking the pH level at regular intervals in their coffee plantation.

pH Meter

Moreover, this pH meter is simple and easy to use. Its reliable, efficient and accurately tests your soil to determine the acidity or alkalinity. The pH scale ranging from 0 to 14 is used to indicate acidity and alkalinity. A pH of 7.0 is neutral, values below 7.0 are acid, and those above are alkaline. The lower the pH the more acid is the soil. The higher the pH, the more alkaline.

The pH values of some common items are- pure water, 7.0- lemon juice, 2.2 to 2.4-orange juice, 3.4 to 4.0-fresh milk, 6.3 to 6.6- mild soap solution, 8.5 to 10.0-most Ontario soils, 4.5 to 8.0.

Determination of the Soil pH

Fertilizer is an important part of growing a productive Coffee plantation. Small vegetable plants are commonly at our back yard, but we are growing it for our selves to meet are own food requirement. So different plants are grown but the need for fertilizer remains the same. The need for fertilizer in a coffee plantation is basic knowledge that we all consider just common sense, however quite a few of us fail to recognize pH of the soil as another factor that is just as important if not MORE important than fertilizer.

pH is more important than fertilizer and here is why. As you know some soils are more fertile than others (more nutrients available). To make our coffee plantation more productive we increase the amount of available nutrients by adding fertilizer. Why & how does pH affect the availability?. Well it gets a little more scientific so lets simplify it a little. The nutrients of a soil are bound up against the individual soil particles. The more acidic the soil the tighter they are bound (as if by a magnet) and hence the less available they are to the plants.

We may be inclined to think that our soils have a good pH just by chance and that we don't need to worry about pH. Well that fact is that with the exception of a few isolated areas in the southern part of INDIA almost all soils are lower in pH than what is needed by most plants to be very productive. Don't forget healthier plants will be more attractive and more productive. Here is why soils are low in pH. One cause has to do with insect and the decay of matter by microorganisms.

This activity over time will increase the acidity of the soil. This causes a big time affect on forest soils. So if your coffee plantations are on soil that was once forest then that will negatively affect pH. We did not realize that many fertilizers with a high first number such as Ammonia Nitrate can actually increase the acidity of our soil. Sure that Ammonia Nitrate give a quick burst of growth but the long term affects if not countered are more acidic soils. Even rain or irrigation causes leaching (removing) of minerals which causes an increase in acidity.

Pertinet facts: You know how you think the grass doesn't grow under your coffee plants very well because you think the tree is taking up all the water. Well just put some lime under those trees and you will be amazed at how well that grass will grow under a tree after the soil is brought back up to a more tolerable level. You see it is the insect and the decay of the leaves that drop the pH under your trees, which is the real culprit!

So now that we know that low pH soils are bad you will want to get the soil pH as high as possible to unleash all those nutrients to the plants. Well sort of although it is fairly uncommon soil can be to high in pH also. Here is why. Bacteria love high pH conditions. If the soil pH gets to high then the bacteria will have a population explosion and they will use up organic matter at a very high rate and can actually deplete the balance needed for plants to grow. To high a pH and micronutrients will become unavailable. Micronutrients affected by pH include iron (Fe), manganese (Mn), zinc (Zn) and copper (Cu).

You can easily increase soil pH by adding lime but if you were to add to much that is something that is not as easily undone. The optimum range is 6.0 to 7.5 Since adjusting your soils pH cost money and since your soil likely has a naturally low pH you will want to bring the pH up to about 6.0 to 6.5 range.

Although there are a million things you can add to the soil to adjust its pH there are probably only two that you will consider and both involve lime. Lime is basically calcium and it will raise the pH of your soil. Lime is commonly bought in two forms, palletized lime and powdered lime also know as agricultural lime.

Powdered lime is by far the cheapest and is applied by a truck that the company you buy it from uses to apply it to your food plot. The other form is palletized lime and it is sold in 50-pound bags and is available at your local hardware store. So access and size of the coffee plantation will likely determine which method you choose.

Here is where things can get complicated if you let them. pH alone doesn't tell you how much lime you need to add. To get true figures you will need to get a soil sample tested. What your soil is made of and how much sand vs. clay is present will alter the amount of lime that is required to raise the pH of a given soil as well as how frequently it will need to be reapplied.

Typically new coffee plantations will initially require a lot of lime per acre to bring the pH up into the desired range. After that smaller amounts of lime can be mixed in with fertilizer to maintain the desired pH. Most colleges with an agricultural dept will perform a soil analysis. These reports will tell you how much lime and fertilizer your soil needs. Or you can purchase your own pH meter and test your soils yourself. Just remember that you can always add more later. So you can forego the soil test if you will take samples test the pH.

Add some lime then retest after the soil has had a couple of months to incorporate the lime. Do this until you get a grasp of how much your lime applications are affecting the soil pH. Remember it is best, even when you have the lab recommended amounts in hand, to apply half of the recommended amount then retest to make sure you don't over do it. Then go back and add more until you bring the pH up to the desired level.

Below are some little important information for lime application

- A clay soil that measures pH 5.5 needs 24 oz of lime a sq. yard to reach pH 6.5
- A sandy soil that measure pH 5.5 needs 8 oz of lime a sq. yard to reach pH 6.5
- Clay soils require more lime initially but less lime to keep it there.
- Dolomite Limestone is slower acting then other forms of lime.
- Lime doesn't travel around your coffee plantation so even application is important.

Dolomite

Most of the plantation crops like coffee, cardamum & pepper growing soils are acidic, the majority of the crop plants produce less than their potential for the following reasons.

— Aluminium,Manganese and Iron Toxicity.

— Calcium, Magnesium, and Molybdenum Deficiency

— Very slow organic matter decomposition

When there are nutrient imbalance in the soil, plants are subject to nutrient stresses that result in reduction in growth, production and quantity. A constant, balanced supply of nutrients to the plant is assential, otherwise nutrient deficiencies or toxicities will result in plant nutrient stresses.

We should test the soil to determine the soil pH, E.C. and which Nutrients are deficient and by how much, nutrient balance in the soil can achieved by proper Fertilizer Management. For this soil analysis is an important diagnostic tool or evaluating and correcting or avoiding the problems. According to the soiled test based recommendation, soil can be neutralized with dolomite (Calcium Magnesium Carbonate)

Uses of Dolomite

— The addition of Dolomite raises the pH, thereby eliminating most major problems of acid soils.

— Toxic soluble Aluminum, Iron and manganese are made insoluble

— Dolomite adds both Calcium and Magnesium.

— Dolomite makes phosphorus in acid more available.

— Dolomite makes Potassium more efficient and economic.

— Dolomite increases the availability of Nitrogen by hastening the decomposition of organic matter.

— Dolomite on acid soils increases available Molybdenum.

pH Paper

At an early age our taste buds indicate to us whether something we eat is sour or bitter. These characteristics of a food can be described as acidic or basic and are due to the "chemical" nature of a substance. Some acids are rather weak and some are very strong. For example, small amounts of weak acids are found in our mouths and stomachs. These acidic solutions serve to help break down and digest the food we consume every day. A weak base is one of the major components found in dishwashing liquid and bath soap.

Acids and bases are important in industry. Industrially, one of the most important acids is sulfuric acid. It is used in petroleum refining, steel processing and fertilizer production. Phosphoric and nitric acid are used in fertilizer production, too. It is important to understand the role of the acidic or basic nature of the soils in which coffee plants are grown. Some plants prefer acidic soils whereas others grow best in basic soil. The acidic or basic nature of the soil can even affect the colour of the leaves and flowers as well as the overall health of the plant.

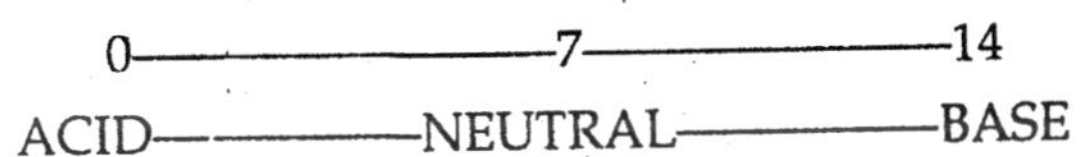

How are pH values determined? Simple. One can use a special pH paper (called Hydrion? pH Paper or pH test paper) which, when placed in a solution, turns a specific colour depending upon the pH value of the substance. The colour of the test strip is matched to a colour chart, which gives the pH value. A pH tester can also be used to measure the pH of soil, water, or other substances.

A pH tester is an instrument that has a probe, which is inserted into a soil or liquid sample and gives a "readout" concerning the pH of the substance tested. Knowing the pH of the soil can help a coffee farmer know what soil type is best.

13

Harvesting and Processing

Harvesting coffee can be done manually or by machine.

Harvesting Methods

There are four different methods of harvesting coffee, the first of the four types being the stripping method. This form of harvesting is done by hand, removes all of the cherries, flowers, green cherries, and deeply over ripened cherries. This method produces poor results because of the mixing of the good cherries with the bad ones. However, it is still practiced in some parts of Africa and Brazil.

The second method uses a comb to brush the trees. This method does remove all ripe cherries leaving the unripe cherries and green leaves still connected to the branches of the tree. This is a time-consuming process that could be worth the time invested, if the plantation owner were being paid a premium for the quality that he was producing. However, this process of harvesting would be more profitable because the unripe cherries will eventually become ripe, increasing the future yield.

The third process used for harvesting is mechanical. This process use a vibrator fixed to the trunk of the tree, shaking the ripe cherries loose so they fall to the ground where they can be reached with ease. The other mechanized harvesting tools are rotating brushes connected to the side of tractors. This process damages the tree ripping off the green cherries, flowers, and leaves at the same time.

The last method is the most expensive because it requires hand picking the cherries when they become ripe. The reason for this expense is that it must be done as many times as necessary until all cherries are picked and in the bins.

Harvest Techniques

Depending on the region where the coffee is being grown, the coffee beans can be harvested as little as once per year to as much as year round, depending on the plant and the climate. Whether or not the plant flowers and fruits is dependent on the cycle of rainy seasons. Growing coffee closer to the Equator gives more and more opportunities to harvest.

The coffee berries are ready 8 to 9 months after the plant flowers. The desired berry is shiny, red and firm to the touch. There are both mechanical and manual ways to harvest the berries. When the berries are harvested by hand, only the ripe berries are chosen, leaving the unripe fruit to be picked later.

Mechanical harvesting is faster and more productive. The machines strip all the berries from the branches at once, regardless of ripeness. This not only damages the trees by taking the berries, small branches and leaves, but means that the ripe and unripe berries have to be sorted out at a later date. Mechanical harvesting can also consist of a shaker that shakes the tree causing the ripe berries to fall to the ground and even tractors with rotating brushes are used to take the berries off the tree, this too can damage the plant.

Harvesting Equipment

The equipment needed for the picking of coffee is pretty simple and inexpensive. The items required include: baskets for the individual picker, holding hooks for bringing branches into position for picking, ladders and containers for transporting large quantities of berries from the orchards to the processing area.

Each basket holds an average of 20 to 25 pounds of harvested berries, and are suspended from the shoulder or fastened with a belt around the picker's waist. The holding hooks are approximately three to four foot-long sticks to which a string or cord is attached. The length of the cord is adjusted to the picker's height, in relation to the average height of the trees. The sticks are usually about 1 inches in diameter at the thickest end. A loop of wire tied onto the cord affords a place for the picker's foot, which can be inserted to hold the hooked branch in place while the picker removes coffee with both hands free.

The picker must be carefully instructed not to bend branches to the breaking point. Ladders are needed for picking when cherries are too high off the ground to be reached with the aid of holding hooks alone. It is recommended that pruning practices which will keep the trees low enough to make ladders unnecessary be used, but when this is not possible, a combination of ladder and hook will usually facilitate picking. The ladders are generally constructed in such a way as to fold for easy carrying, with a center hinge inserted in a divided top crosspiece so that the picker's weight will hold the ladder rigidly in place. This type of ladder seems to be one of the safest, especially since coffee is often grown in rough and irregular terrain.

Coffee Crop Maintenance

Maintaining a coffee crop, outside of planting and harvesting, is not as involved as other activities in processing coffee. Unless the crop is in an area where annual rainfall must be augmented by irrigation, there are no irrigation needs for the crop. The usual weed and pest control measures are in place as they are with any agricultural operation. Fertilizer is applied to ensure the crop receives the proper required nutrients.

One of the biggest activities in a maintenance program is pruning the coffee trees. Pruning is performed for several reasons. The first reason is to maintain the physical size and appearance of the bush to allow for ease of maintenance and to allow for harvest activities. Second is to encourage productive growth and to keep the canopy at the right density. Deadwood and branches do not produce coffee berries. A canopy that is too thick can choke out the productive parts of the bush located under the plant.

Careful selection of red cherries at harvesting is essential for good quality coffee. To make pulping and grading easier, process only ripe, red cherries; do not use a mixture of red, over or under-ripe cherries. Potential damage to coffee beans is reduced as the pulping machine can be better adjusted to the one type of red cherry.

In Lao, harvesting for Arabica takes place from October through December and for Robusta, December through February. Clean, washed bags should be used to collect the harvested fruits; NEVER use bags that have contained fertiliser or other chemicals. Cherries should be processed the same day as harvesting and should not be mixed with the previous day's harvest. Equipment and sorting areas should be checked daily and kept thoroughly washed clean. Any fermented part of cherry from the previous day will contaminate the newly harvested cherries and result in deteriotation of the entire batch. Carefully wash and sort cherries before starting the processing to remove twigs, leaves or other foreign matter.

Methods of Coffee Processing

Coffee processing transforms fresh coffee cherries into clean, green bean of 12% moisture ready for export or for roasting. This process involves harvesting, pulping, fermenting, washing, drying, hulling, cleaning, grading, sorting, storing and transporting green beans. The process can be broadly divided into two main components - Wet Processing (cherry to dry parchment) and Dry Processing (dry parchment to exportable green bean).

It is important to understand that each of these steps has an influence on the final quality of coffee produced. Processing is a chain of activities aimed at achieving a coffee of high quality. If any link in the chain is broken (such as over-fermentation, mould contamination, taints or odours or physical damage to the bean) then that loss in quality can never be regained.

Three main processing methods are the basis for the range of coffee processing techniques used throughout the world - natural, semi-wash and full-wash (Figure 1 shows the last two processes)

Natural Process

This is a one-step operation where the coffee bean is dried inside the whole coffee fruit to 12% moisture. The dry cherry is then hulled to produce a dry green bean. This is the low cost, traditional system resulting in a low quality coffee, and is not recommended. In Lao, most Robusta coffee is currently processed this way.

Full-washed Process

The skin of the fresh cherry is physically removed using a pulper machine with addition of water (pulping).

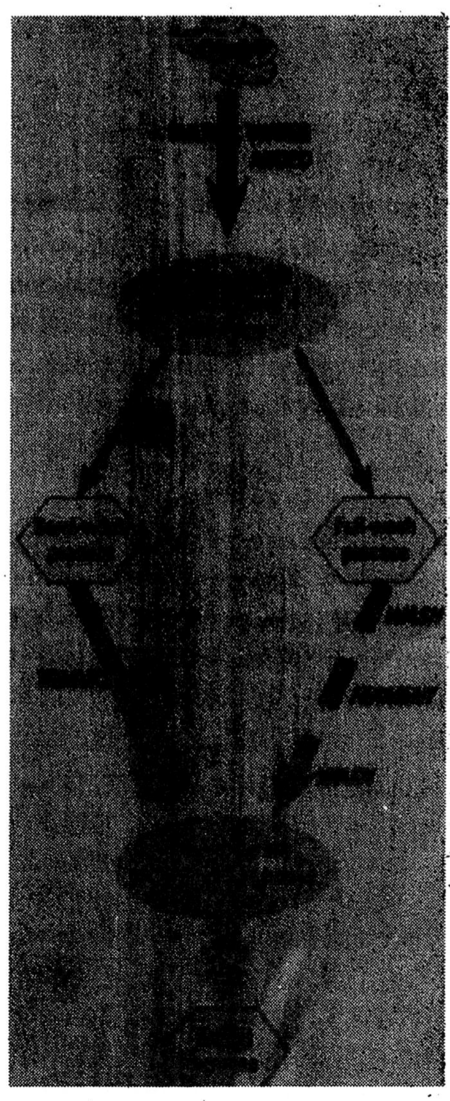

Figure 1. Simplified diagram of semi and full washing processes

The sugar coating (mucilage) is allowed to ferment over one to two days and then the parchment is washed thoroughly to remove all traces of fermented mucilage. The parchment is dried until the bean inside reaches 12% moisture. This process can produce high quality coffee, but requires large quantities of water and requires very good management of the fermentation and washing process to ensure the coffee flavour is not damaged in the process.

Semi-washed Process

The skin of the fresh cherry is physically removed by a pulper machine with addition of water, as with full-washed processing. The mucilage is then removed immediately after pulping using a demucilager. Notably this process does not ferment the mucilage as it is mechanically removed by a demucilager. Immediately after demucilaging, the clean parchment is ready for drying until the bean inside reaches 12% moisture.

Recent studies in Lao and Myanmar have shown that pulper/demucilager units are a cost efficient and an effective way to consistently produce high quality coffee without the need for fermentation and washing. These units typically use only 0.5 L of water per kg of fresh cherry and reduce the risk of over-fermentation and quality problems in the final coffee product. While there is an initial capital cost to purchase the pulper and demucialger units, there is no need for fermentation tanks and washing systems. Pulper/demucilager units are recommended for semi-washed wet coffee processing, in Lao.

Natural Process

Coffee cherries are laid out in the sun to dry.

Semi-wash Process

Pulper and demucilager units produce clean parchment coffee ready for drying. Inexpensive and good for smallholders; processes 0.5 MT/hr cherry.

Full-wash process	*Full-wash process*
Pulper unit removes skin	Cherry is washed, fermented and washed again to remove the mucilage

Drying Process

Drying can be done in full sun on a hard, flat, clean surface such as concrete slabs, tarpaulins, mats, raised tables or trays with a mesh base. Drying should remove moisture from the coffee bean in a slow and continuous process until the bean is at 12% moisture. Drying coffee directly on soil or dirty surfaces can lead to dirty or earthy flavours in the finished coffee. Rewetting of the coffee or storage of partially dried coffee due to rain is a major problem facing sun-dried coffee. Drying coffee too slowly by spreading it too thick on drying areas is also a major problem. Each of these situations can lead to fermented or fruity flavours in the coffee along with mould-growth producing mouldy or musty flavours.

Controlling the drying process to ensure that coffee is not over-dried is important. Over-dried coffee is easily damaged during hulling and may also result in a bland flavour in the final cup. Drying cherry coffee may take 18 to 20 days. Parchment coffee dries in about 9 to 10 days. During the precess, coffee must be covered with polythene or plastic sheets if rain occurs and every night to stop re-wetting that results in mould development. Coffee is fully dry when green bean is a translucent, jade green colour and 12% moisture content. When bitten with the teeth, the bean is dry when it is barely marked, and over-dry (8 to 10% moisture) if it breaks.

Storage of Dry Parchment

Once parchment has been dried so that the green bean has reached 12% moisture, it can be stored while the grower / processor decides when it will be sold or hulled. Mould can grow on stored coffee if it has not been dried sufficiently before storage or if the stored coffee absorbs moisture from the atmosphere due to humid conditions. This can lead to mouldy or musty flavours. Storage areas must be kept isolated from strong smelling liquid such as petrol or diesel, or agricultural fertilizers and chemicals, as stored coffee can take on these odours which will continue to the final cup.

Parchment coffee or dry cherry is stored on-farm in either jute bags sometimes covered with polyethylene covers, or in woven polyethylene sacks covered with a polyethylene sheet, or in special polyethylene bags or silos. If not carefully managed, parchment or green bean stored in uncovered jute sacks in a moist climate, will absorb moisture and go mouldy. Poorly ventilated warehouses and relative humidity situations over 65% will create mould problems.

Hulling and Sorting Dry Parchment

Hulling dry parchment is a mechanical process to remove the dry parchment skin and silver skin from the green bean (Figure 2). If the huller is set incorrectly or the coffee is over-dry and brittle, coffee beans can be damaged. If the coffee is too wet the beans can be crushed. There are a range of machines that are able to clean and sort hulled coffee by colour, size, density and aerodynamic shape. Ultimately the human eye is used as the final process to 'hand-sort' coffee ready for export. However, even with the wide range of machinery available, coffee that has picked up off-flavours but otherwise looks normal, cannot be sorted, and is only identified in the cup when it is too late.

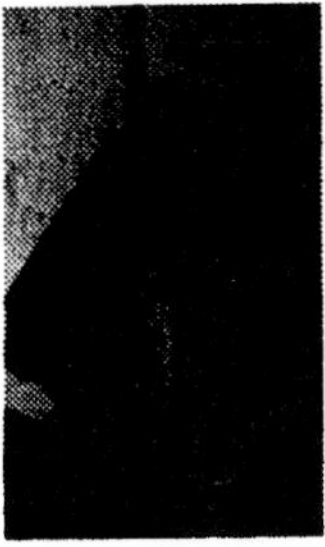

Figure 2. Hulling machine (above) and beans with parchment removed after hulling

Storage and Shipment

Stored, green bean is very susceptible to being contaminated by nearby chemicals or fuels. Storage and shipment of green bean in jute sacks that have been made on machinery lubricated with petroleum oils, can lead to a 'baggy' or 'oily' taste in the coffee. Use clean, jute sacks specially made for coffee.

Green bean that is stored for long periods in hot and humid conditions is liable to absorb moisture from the atmosphere with resultant mould producing musty flavours. To ensure minimum spoilage, beans in jute sacks or woven poly bags should be evenly stacked in a well-ventilated area that remains at less than 65% relative humidity. After some time in storage, the bean surface begins to oxidise leading to 'woody' taints. Coffee should not be stored for longer than 12 months as the beans fade and mottle.

Transport

Storage and transport pose similar risks to coffee quality.

Re-wetting of beans due to leaky tarpaulins, or high humidity inside hot containers standing for long periods in tropical ports, can result in the coffee developing mouldy or musty flavours. Special techniques for handling bulk or bagged green beans for container shipping are now well known.

14

Dimensions of Coffee Production

There are four themes intimately related to the coffee sector: biodiversity and conservation of forest ecosystems; agro-chemical use; water pollution from coffee processing; and soil quality.

Biodiversity and Conservation of Forest Ecosystems

Deforestation trends are serious throughout the coffee-producing lands of Latin America. Seven of the ten countries in the world with the highest deforestation rates are in Latin America and the Caribbean; these seven countries include Jamaica, Haiti, Costa Rica, Paraguay, Ecuador, Guatemala and Mexico. In a number of areas, tropical forest ecosystems have disappeared or are on a path to elimination in the near-term. By the late 1980s, for example, only an estimated one-fourth of the primary moist tropical forest in Colombia remained.

Remarkable biodiversity values are at stake. Latin American tropical forests are critical ecologically for purposes of protection of atmospheric dynamics, water quality, and wildlife species, as well as economically as reservoirs of germplasm with multiple applications for food, medicine, and industrial products.

The region's threatened natural heritage transcends national boundaries. For instance, neotropical migratory birds that winter in northern Latin America constitute 60 to 80 percent of the bird species that inhabit forests throughout the eastern U.S. and Canada; neotropical migrants also constitute a large fraction of bird species in the forests of the Pacific Northwest. Birds numbering in the hundreds of millions and representing more than 120 species migrate annually through or to the part of the Central American isthmus composed of Costa Rica and Panama.

Traditional, shade coffee production has been shown to be highly beneficial to biodiversity conservation in tropical forest ecosystems. In northern Latin America, traditional coffee covers very significant areas with closed canopy, agro-forestry systems with high species diversity. Smithsonian Migratory Bird Center biologists conducting research in the southern Mexican state of Chiapas discovered that traditionally-managed coffee and cacao (chocolate) plantations support at least 180 species of birds, an amount significantly greater than bird numbers found on other agricultural lands and exceeded only by undisturbed tropical forest. The attraction of industrial sun coffee for birds falls well short of that seen in traditional shade systems. For example, studies in Colombia and Mexico have identified over 90 percent fewer bird species in sun-grown plantations than in shade coffee.

Shade coffee also provides essential habitat for diverse communities of other tropical forest species. Findings by University of Michigan biologist Ivette Perfecto and colleagues from research in Costa Rica suggest that local species diversity of beetles, ants, wasps and spiders on a single tree species (*Erythrina poeppigiana*) in shade coffee plantations approximates the arthropod diversity levels on single tree species sampled in undisturbed tropical forest.

Additional recent studies on tropical forest ecology have been conducted by scientists from Mexico's National University and Chicago's Lincoln Park Zoo. These researchers' work in Veracruz, Mexico, has shown that shaded agricultural plantations, as compared to unshaded agricultural landscapes, feature richer diversity of small mammals such as opossums, squirrels and mice. Bats, important dispersers of seeds and pollinators of many tree species, as well as natural predators of insects, also show a presence in such systems. Comparing forest habitat to several agricultural lands, these same researchers found that habitats designated as "mixed plantation" (cacao, coffee, bananas, and citrus) and "coffee" (coffee with shade trees) jointly contained 74% of the species richness.

Traditional coffee is often integral to agro-forestry systems in which tree species are cultivated together with the coffee and other agricultural commodities. Where geographic and market conditions are favorable, economic returns can be achieved through sustained-yield timber production in association with coffee. For example, research in Costa Rica has shown that timber from the precious hardwood species *Cordia alliadora* can occur with no significant damage to growing coffee crops. Agro-forestry systems, including those involving coffee, have potential to enhance the economic and ecological stability of poor rural areas in northern Latin America. By providing an alternative to deforestation, traditional coffee systems constitute an important check against greenhouse gas emissions that contribute to global warming.

Agrochemical Use

Traditional shade coffee systems typically rely on much lower chemical inputs than industrial plantations. This is because planting coffee among natural vegetation, or among trees planted for shade, fruit or timber, can reduce susceptibility to pests. Moreover, because many traditional methods have been passed down to today's farmers by previous generations before synthetic pesticides and fertilisers were widely used in agriculture, a human-land use equilibrium has evolved in coffee production over time.

Intensive pesticide use within industrial coffee production often employs chemicals that present serious health and ecological concerns. Sampling of imported green coffee beans conducted by the U.S. Food and Drug Administration (FDA) in the late 1970s and early 1980s revealed frequent detections of DDT, BHC (benzine hexachloride) and other pesticides banned in the U.S. because of possible carcinogenicity or long-term persistence in the environment. In 1983, the Natural Resources Defense Council retained the services of an outside contract laboratory to conduct independent testing on imported coffee beans. The analysis revealed multiple pesticide residues on all samples when green coffee beans were tested using detection methods many times more precise than the FDA procedures.

The roasting process reduced detectable levels of pesticide residues on the bean samples; however, the test of one sample of the Brazilian coffee beans retained the original level of DDD (the toxic metabolite of DDT) that had been detected on the beans before roasting. It should be noted that while DDT is rarely used on coffee today, other chemicals are used to combat insect pests, weeds, and diseases.

Over the last decade, governments throughout the Western Hemisphere have taken steps to prohibit use of a number of pesticides banned in the U.S. Certain banned chemicals remain approved for agricultural use in some Latin American countries, however. For instance, a 1990 report from the General Accounting Office found that Costa Rica continues to permit use of chlordane, a highly toxic insecticide that persists for years in the environment. Attempts to restrict U.S. exports of banned pesticides have failed in recent years; for example, "circle of poison" legislation passed by the U.S. House and Senate fell short of final enactment in the 1990 farm bill.

Under-regulated pesticide use also threatens farmers and other rural residents with exposures to toxic substances in the workplace or in water supplies. For example, serious public health and water quality impacts have been linked to pesticide use in Mexico; in one documented case in 1987, more than 200 people became sick from drinking water contaminated with agricultural pesticides and fertilisers in the western Mexican state of Jalisco.

A recent World Resources Institute (WRI) report documented extensive human exposure to pesticides in Latin America and elsewhere in the developing world; for example, studies of farmworkers and their families in Nicaragua have revealed significant decreases in the activity of cholinesterase, an enzyme vital for normal neuro-muscular functioning. The WRI report notes that "inadequate safety and hygiene practices are the norm" in developing country pesticide use.

Recently, concerns have been raised about human health and environmental impacts associated with expanded use of the highly toxic insecticide, endosulfan, in Colombia to combat a coffee insect pest known as "la broca." According to official accounts compiled by Pesticide Action Network North America (PANNA), more than 100 human poisonings and one death were attributed to endosulfan use in coffee during 1993; more than 100 poisonings and three deaths were reported in 1994. Although the Colombian health ministry took steps to ban endosulfan use in January 1995, concerns continue to be raised that this move has not been implemented fully. The Colombian Coffee Growers' Federation and the country's National Center for Coffee Research have pointed out the availability of less-toxic chemicals, as well as biological methods for coffee pest management, and have prohibited their field technicians from recommending use of extremely or highly toxic pesticides on coffee.

Farmworkers historically unaccustomed to technified coffee production systems encounter an array of chemicals that are supposed to be applied with protective gear such as masks, long-sleeved shirts, long pants and boots — clothing that frequently goes unused in the heat and humidity of tropical environments. Recent field research by the U.S. Environmental Protection Agency (EPA) documented widespread lapses in the use of protective clothing among pesticide applicators working in the Mississippi Delta. These EPA staff findings have important implications for farmworker health and safety in technified coffee production in northern Latin America. Current pesticide regulatory systems, including in the United States, take insufficient account of the practical limitations of wearing protective gear in hot weather.

Increased nitrogen fertiliser applications have gone hand in hand with the widespread removal of shade cover from Central American coffee plantations. Heavy synthetic fertiliser inputs in coffee have contributed to nitrate contamination of drinking water aquifers in Costa Rica, with the documented groundwater pollution in some cases exceeding World Health Organisation levels. In high concentrations, nitrates can cause infant methemoglobinemia ("blue-baby syndrome"), a potentially fatal condition that impedes oxygen transport in infants' bloodstreams. Other human health concerns surrounding nitrate contamination of groundwater include suspected links between nitrates and certain cancers, birth defects, hypertension, and developmental problems in children.

Water Pollution

Largely irrespective of how coffee is grown, discharges from coffee *beneficios* (processing plants) represent a major source of river pollution in northern Latin America. The process of separating the commercial product (the beans) from coffee cherries generates enormous volumes of waste material in the form of pulp, residual water and parchment. For example, the Guatemala-based Instituto Centroamericano de Investigación y Tecnología Industrial estimated that over a six month period during 1988, the processing of 547,000 tons of coffee in Central America generated 1.1 million tons of pulp and polluted 110,000 cubic meters of water per day, resulting in discharges to the region's waterways equivalent to raw sewage dumping from a city of four million people.

Coffee *beneficios* exist in a wide range of sizes. In Guatemala, for instance, where a total of some 4000 processing facilities are estimated to dot the landscape, the National Association of Coffee Growers divides them into micro-facilities (those with a capacity to process 500 to 5000 pounds of harvested coffee per day), medium facilities (5000 to 50,000 pounds per day), and large (greater than 50,000 pounds per day). The 100 *beneficios* belonging to this last category (3% of all Guatemala's *beneficios*) process 60% of the coffee produced annually

Ecological impacts result from the discharge of organic pollutants from *beneficios* to waterways, robbing aquatic plants and wildlife of essential oxygen. Costa Rican health officials have expressed concerns over harms to marine life along parts of the Pacific Coast where rivers contaminated by coffee processing wastes flow into the ocean. According to Costa Rican government estimates from the early 1980s, coffee processing residues account for two-thirds of the total biochemical oxygen demand (the principal measure of organic pollutant discharges) in the country's rivers. In 1992, Costa Rica instituted a plan to upgrade the nation's coffee processing systems, with the objective of cutting organic pollutant discharges to surface waters by 80 percent within five years.

Recent years have witnessed important progress in the development of pollution control technology in coffee processing. A small but growing number of *beneficios* are substantially reducing the volume of water used in "wet" processing of coffee; this in turn, reduces the amount of water requiring treatment before being discharged from the processing facilities. Additional environmentally sound measures include composting coffee husks mixed with farm animal manure to use as organic fertiliser on crops, as well as digesters that produce methane gas that can be used for practical applications like powering the processing plant. Success has been demonstrated with these measures in various parts of northern Latin America, including the major coffee areas of Mexico's Veracruz state. Without a concerted regional investment plan in improved technology, however, pollution prevention will remain the exception to the rule in this part of Latin America.

Soil Quality

Soil quality benefits of traditional agricultural resource management in northern Latin America have been well documented by Colorado State University geographer Gene C. Wilken in his 1987 book, *Good Farmers*. Regarding soils, Wilken attributes the success of traditional systems to key factors such as the following:

- — understanding of local resource characteristics and how they can be applied efficiently to particular soil and crop conditions to build organic matter through such practices as mulching and multiple cropping;
- — careful, precise applications of inorganic nutrients to save on production costs and prevent damage to soil and water quality; and
- — effective slope management techniques such as terracing that conserve moisture, control erosion and enable agricultural production in areas otherwise unsuited for farming.

Such practices are typical within shade coffee production systems, which demonstrate remarkable technical sophistication in soil management. Additional benefits derive from substantially reduced or foregone use of pesticides, whose over-application can eliminate insects and micro-organisms that play vital roles in the enhancement of soil productivity and plant nutrition.

Elimination of shade cover can cause significant impacts on various soil quality parameters. Research in Nicaragua in the late 1980s documented that, relative to traditional systems, significantly higher erosion rates occurred on renovated coffee plantations where shade had been reduced.

Nutrient cycling also reacts to changes in the shade cover in coffee. In Costa Rica's Central Valley, where rainfall can reach up to 2.5 meters annually, the leaching of soil nutrients into the groundwater can be significant. Within these high-rainfall areas, unshaded coffee loses nearly three times more soil nitrogen than shaded plantations. In general, shade coffee systems have been shown to be more conservative recyclers of nitrogen than unshaded plantations.

15

Diversification Forestry in Coffee-Producing Countries

The world's coffee soils are generally friable and loamy, of lateritic or volcanic origin and generally brown, chocolate or red in colour in Brazil, *terra roxa* or red soils are always favoured for coffee cultivation. The two chief commercial species, Arabica (*Coffea arabica*) and Robusta (*Coffea canephora*) and their varieties grow in somewhat different environmental conditions which, however, generally occur within the vegetation formation of tropical humid forests.

Arabica coffee is usually associated with the tropical humid forest type at high altitudes. In its natural habitat in Ethiopia, its normal altitude is between 1,800 and 2,400 meters above sea level. In other countries the natural environmental conditions of climate can be simulated by a suitable combination of latitude and attitudinal zonation, provided the essential humid character of the forest type remains unchanged and a suitable range of edaphic conditions obtains.

Robusta coffee is indigenous to African equatorial ram forests from the west coast to Uganda and southern Sudan. It occurs from sea level up to 1,200 meters and is found in a natural state under dense shade. As a rule, therefore, Robusta has been introduced under warmer humid conditions, and Arabica under more temperate cooler conditions. For Robusta, it is essential to retain natural or introduce artificial shade in conformity with the natural forest environment.

For Arabica, the question of shade is more complex. Shade is generally considered necessary, except under a combination of especially favorable climatic and edaphic conditions. In the same latitudinal range and with altitudes generally above 1,500 meters, with cooler temperatures, more evenly spread rainfall and greater humidity, shade appears to be unnecessary. Such areas occur in the Kenya highlands above 1,500 meters. Similar areas are also found in climatically and edaphically similar but more temperate zones, for example in most of the southern states of Brazil

Forestry in Coffee Areas

Arabica

In the Arabica areas, both in Africa and America, the coffee-growing zones usually have definite forestry interests, particularly from the point of view of their present and/or potential site productivity, in developing industrial plantation of fast-growing species, chiefly conifers and eucalypts. In Africa, however, although there is considerable overlapping between the forest and coffee-growing areas, the main planned expansion of forest plantations lies outside the coffee-growing areas. From the standpoint of land use, coffee plantations in Africa are generally well maintained and not particularly subject to erosion or other hazards, largely as a result of government policies which have established strict standards for coffee growing and are supported by good technical extension services. The position in Latin Americana varies between individual countries.

In Brazil, the main timber and coffee-producing areas overlap, chiefly in the southern and eastern regions of the Paraná pine forests which provide 60 percent of the country's industrial wood supplies. The pine resources in the State of Paraná dwindled by 1963 to an estimated area of 1.5 million hectares, out of a total forest area of 6.5 million hectares. Between 1953 and 1963 the annual clearing of forests for agricultural crops in Paraná was estimated at 270,000 hectares, of which 250,000 were tropical or subtropical broadleaved forest and the remainder Paraná pine (*Araucaria angustifolia*). The total volume of *Araucaria* in Paraná was estimated in 1963 at 50-60 million cubic meters and the annual cut at 4 million cubic meters.

At this rate of cutting, the Paraná pine forest resources might be depleted of saw timber by 1975-80. However, the actual coffee-growing areas are concentrated mainly in the tropical broadleaved forest zone. An important consideration is the incidence of frost in parts of this zone, which causes severe damage to coffee. Frost is also a major contributory cause of fires in seasons of drought which spread throughout commercial forests and plantations. For instance, about 2 million hectares of private forests in Paraná were swept by fire in 1963, including about 20,000 hectares of industrial plantations.

Extensive deforestation of commercially valuable forest in some of the coffee-growing states, for example in the State of Minas Gerais for industrial fuelwood supplies, has created shortages in timber supplies. Erosion caused by deforestation with consequent landslides and floods, is also serious in some areas around Rio de Janeiro.

In Colombia, the main potentially productive forest resources and also the potential sites for industrial plantations are outside the coffee-growing areas. The central range of the Andes, the main coffee zone, originally supported dense montage forests, now replaced by coffee plantations. Well-managed coffee maintains and protects the soil conditions adequately, and has proved to be the most profitable agricultural crop.

Replacing well-managed coffee with forest crops is usually neither desirable nor practicable, although there is some scope for reforestation to control erosion. Natural forest is now largely above 2,500 meters, the maximum altitude suitable for coffee here reforestation for both protective and productive reasons is necessary, and it has already been carried out on a small scale with some measure of success. The native bamboo (*Guadua angustifolia*) is associated with the coffee zone, occurring chiefly as relict stands on land cleared for coffee or pasture at altitudes of 900 to 2,000 meters. It is the principal source of construction material and serves a variety of agricultural uses in the coffee areas. It has also been found acceptable up to 20 to 25 percent of the mix in the production of kraft paper by one of the principal pulp and paper plants.

In Mexico and Central America, the coffee zones lie outside the main productive or potentially productive forest areas. The subtropical wet forest type is associated with the major part of the coffee zones in this region at altitudes of 300 to 2,000 meters. There are no important timber species in these forests, though some medium-value timbers occur, which provide wood for local uses. Subtropical natural coniferous forests, chiefly of *Pinus* spp. Occur generally at altitudes above the coffee zone in Mexico and Guatemala.

In Mexico, about half of the coniferous forests is concentrated in the states of Durango and Chihuahua, in neither of which coffee is grown. However, some important coniferous forests are also found in the coffee-producing states of Oaxaca, Chiapas, Puebla and Hidalgo. Although there is a potential for planting fast-growing species, both inside and outside the coffee-growing zones, there are hardly any commercial plantations. The major exceptions are plantations in the region of Valles, which supply the raw material for the production of.

Guatemala's total pine forest area has been recently estimated at 2.S million hectares, of which the pine forests of Sierra de las Minas and Chuaous are estimated to cover 500,000 hectares and to contain some 50 million cubic meters of wood. These figures suggest that Guatemala has a very large volume of pine, much larger than that indicated by an estimate of a recent IBRD mission, thus placing Guatemala in almost the same rank as Honduras in pine resources. El Salvador has only a limited area of pine forests in the north.

Robusta

The main Robusta growing countries - Ivory Coast, Angola, Uganda, Tanzania and Congo (Kinshasa) - lie in the equatorial forest belt which is classified as humid to subhumid, mainly semideciduous forest at low and medium altitudes, with an average annual rainfall of 1,500 to 2,000 millimetres, occurring on flat or slightly undulating land. This type of forest extends in Angola to the *dembos,* or coffee forests at 400 to 1,000 meters in altitude on very rugged topography. The equatorial rain forests in west and central Africa are associated with the Robusta coffee-growing areas and play a significant economic role by forming the basis for an organised and well-established trade in timbers of high commercial value which are exported mainly to Europe.

Ivory Coast is a particularly important producer and exporter of hardwoods. The semideciduous forest type is the most productive, containing the main species of redwoods such as the mahoganies (*Khaya and Entandrophragma*), *Mimusops* and *Afromosia* and general utility species such as Obeche or Wawa (*Triplochiton scleroxylon*). The semideciduous forest, found under somewhat drier conditions as gallery forests in the savanna, also contains species such as Iroko (*Chlorophora excelsa*) of commercial value. The coffee zone extends marginally to such gallery forest. In recent years the area of closed forest in Ivory Coast has shrunk from an estimated 13 million to about 7 million hectares of which only 2.8 million are reserved forests. Continued encroachment by shifting cultivation, generally following exploitation and new plantations of agricultural crops, is reducing the forest area by about 120,000 hectares annually, in particular along the periphery of roads newly opened for the extraction of forest products. The area under coffee increased from 385,000 to 720,000 hectares in the period 1959-66.

Simultaneously, commercial timber fellings increased rapidly under the impact of export demand. Felling wood for processing increased from 320,000 cubic meters in 1955 to 2.55 million cubic meters in 1965, 85 percent of which was exported, mainly as logs. Annual timber exports grew from 1,100 million CFA francs in 1953-55 to 15,700 million CFA francs in 1963-65- that is, to the second place after coffee. The combination of progressive alienation of forest land and the increasingly large annual cut of timber is exhausting the country's resources of woods now in demand which, at the present rate of depletion, are likely to be exhausted in about 30 years. There is, therefore, an immediate need for halting the expansion of agriculture into productive forest land.

Growth of Forestry and Forest Industries

In Brazil, the main coffee-growing states, São Paulo, Paraná and, to a smaller extent, Minas Gerais and Santa Catarina, are the most industrialised. Forest industries are either directly wood-based or linked with other major industries, for example with metallurgical industries in Minas Gerais. Of primary wood-using industries, the most important is sawmilling, which is largely based on the Paraná pine forests. Plywood and veneer are also of importance, especially in the State of Paraná In the four States mentioned, annual production in 1961-65 was about 3 million cubic meters of sawtimber of pine and 0.7 million cubic meters of timber of other species.

Production has decreased slightly since 1961, indicating that the resources of Paraná pine are being strained to keep up with production. Paraná pine accounts for most of the timber exports it provided annually 1.9 million cubic meters in 1964-66 valued at U.S.$51.8 million, compared with the total production of 2.2 million cubic meters and the value of $70.1 million. Industrial charcoal is especially important in the State of Minas Gerais where the natural forest has been severely overexploited. In 1963-65, annual production was 728,300 tons, valued at 7.73 million cruzeiros. Industrial plantations in Minas Gerais supplying this industry now cover about 50,000 hectares.

The largest forest-based industry in Brazil is that of pulp and paper, particularly in São Paulo and Paraná but it is also expanding in the two southernmost States. This industry depends largely on natural Paraná pine from private forests and plantations, of which São Paulo possesses the largest extent. Many of the principal enterprises own large plantations of eucalypts and some softwood plantations (*Pinus elliottii* and *P. taeda* mainly) for the supply of long-fibered pulp are expanding.

These enterprises employ the most modern mechanised techniques in planting, some supported by comprehensive experimental research centers. As a result of this progress, Brazil has in recent years become practically self-supporting in pulpwood and paper. Prior to 1965, the country relied strongly on imports of pulp for its production of paper and paperboard but a deficit of 51,965 metric tons of pulp in 1964 was transformed into a surplus of 27,750 metric tons in 1965 in 1966, there was only a small pulp deficit (940 metric tons). Recorded apparent consumption of paper and paperboard rose from 311,000 metric tons annually in 1949-51 to 713,000 metric tons in 1963.

In Colombia, the principal forest resources and forest industries are outside the coffee zone. Annual exports of round and processed timber are estimated at about 200,000 cubic meters valued at U.S.$4 million. The current trend is for a rapid increase in processed wood exports, especially of sawn timber, which in 1966/67 exceeded the exports of round timber. With the recent development of pulp and paper plants, exports of paper and paperboard increased to 3,600 tons in 1965. Imports of timber and wood products amount to less than $0.5 million per year.

Developments in paper and paperboard have constituted the most dynamic element in the forest products sector consumption increased from 73,000 to 240,000 tons from 1958-66, and local production expanded even more rapidly. The raw material for paper and paperboard production is chiefly from tropical hardwoods in the Southwest Pacific region. Plantations supply mostly local fuelwood and pole requirements and contribute little to industry. In the Department of Caldas, one of the chief coffee-growing areas, some 800 hectares of plantations existed in 1967, mainly as protection for the reforestation of degraded pastures or the improvement of catchment areas. The main contribution of forestry in the coffee zone is made by the indigenous bamboo (*Guadua angustifolia*).

In 1964 bamboo to the value of 1 million pesos was sold at Caldas timber market compared with 13 million pesos for fuelwood and charcoal this hardly reflects the importance of the species in the economy of the region. The future consumption for industrial use has to be re-examined, since it is already accepted up to 20 to 25 percent of the pulpwood mix in one of the biggest plants in Cali. The excellent growth and extensive use emphasize the need for a survey of existing *Guadua* bamboo forest and of its actual and potential contribution to forestry development, including its protective role in the area.

Among the Central American coffee-producing countries, El Salvador is the least important in forest industries, since it lacks substantial forest resources. The country has only about 226,000 hectares of forest, out of which about 30,000 hectares are conifers with a growing stock of about 1.4 million cubic meters. Sawnwood production in 1964 was estimated at only 6,000 cubic meters and consumption at 60,000 cubic meters. El Salvador imports approximately 90 percent of its requirements, mainly from Honduras. The consumption of pulp and paper products is comparatively high, as reflected by the 1964 figure of 28,700 metric tons.

In Guatemala, the main industrial production and export potential lie outside the chief coffee-growing zone. Guatemala has a net deficit in wood and wood products, largely because its paper and paperboard consumption is increasing. Fuelwood is still the main product. Production of industrial wood, mainly from coniferous forest resources, is underdeveloped it amounted to an annual average of only 824,000 cubic meters in 1961-63. Sawmilling average production was 100,0 cubic meters annually in 1964-65, plywood 3,000 cubic meters in 1965, and particle board 2,400 metric tons in 1967. No pulpwood mills exist.

About 8,500 metric tons of paper and paperboard products are currently manufactured annually from a mill based on domestic nonwood fibered pulps, waste paper and imported pulp. There is a vast potential for expanded industrial production in the north, chiefly in the Department of Petén. Recent information on the coniferous wood resources prevalent in the coffee zone indicates that, with the extensive pine forest resources of Guatemala, serious attention should be given to the future role of Guatemala in pulp and paper production to meet the rapid growth in consumption in Central America.

Mexico has rich forest resources. However, although fairly well advanced, forestry does not make a significant contribution to the national economy, and generates only about 0.3 percent of the gross domestic product (GDP), decreased from an estimated 0.5 percent in 1950. Forest industries contribute some 4 percent of GDP. The value of the pulp and paper output increased from U.S.$88 million in 1958 to $128 million in 1963. Approximately 70,000 persons are directly employed in forestry and forest industrial activities. Under the current system of *unidades*, exploitation units are given out for specific industrial needs. Government control and jurisdiction is maintained in these units, even on private lands. Some of the major coffee-growing States are also important contributors to forest production and industry, chiefly based on natural pine resources.

In Angola, forestry and forest industries have developed largely outside the Robusta coffee zone. However, the relatively small humid forest in the coffee-growing Cabinda district provides one third of Angola's timber, contributing mostly logs and some processed timber to exports. About one half of Angola's total wood supplies comes from the Miombo *Brachystegia* woodlands. In the highlands of the central plateau which already includes small areas of Arabica coffee, mainly private interests have established highly successful Eucalyptus, some cypress (*Cupressus lusitanica*) and pine (*Pinus patula*) plantations totaling 103,000 hectares in 1965.

These plantations now supply most of the fuelwood requirements of the Benguela Railroad Company and, since 1964, pulpwood to a pulp and paper mill at Alto Cadumbela. At present, the long transport distances and the lack of cheap long-fibered raw material are serious drawbacks. However, an almost twofold expansion of production was planned for 1968, depending on the possibility of increasing the plantation area within an economic range of transport.

Ethiopia has approximately 4 million hectares of closed forests and about the same area of unproductive forest, woodland and some bamboo forests. However, the forestry contribution to the national economy is small, and it is heavily weighted by fuelwood. Private plantations of *Eucalyptus globulus* (16,000 hectares), mainly found in the neighbourhood of Addis Ababa, supply some poles as well as fuelwood. Ethiopia is a net importer of forest products and has no exports. Imports are mainly of paper and paperboard (70 percent) and wood manufactures (26 percent). The average annual value of imports in 1961-65 was Eth. $5.78 million. Timber industries are poorly developed and comprise some sawmills, one plywood factory and one new particle board factory.

In Ivory Coast the strongest development in forest industries allied with expanding exports of high-value woods, has taken place in the Robusta coffee zones. With the exception of the inaccessible southwest region, the coffee zones coincide with the zone of commercially productive humid forest. In 1965 concessions covering a surface area of about 6.7 million hectares were in force and, despite the government decision to stabilise production at the 1962/63 level, the output of industrial wood in 1965 rose to nearly 2.9 million cubic meters. Although a considerable extent of forest area was lost to agriculture, the size of compensatory plantations has been negligible, the total to-date being 13,720 hectares, mainly 6,000 of teak and 5,700 of cashew (*Anacardium*).

The current trend in production concentrates largely on timber in log form for export. The value of log exports increased over 166 percent between 1960 and 1964. There was also an increase in sawnwood production from 157,000 cubic meters in 1963 to 258,000 cubic meters in 1965, and diversification into new veneer and plywood industries producing approximately 10,000 cubic meters each in 1965. The situation is likely to improve in the near future because of:

1. the recent inventory of the forests in the southwest region, which indicates that wood resources are substantially larger than estimated earlier, with the possibility of stabilising log removal at about 3

million cubic meters per year and proportionately increasing the use of secondary species

2. the planned establishment of about 58,000 hectares of industrial plantations. There is no pulp and paper production at present in Ivory Coast. While the average annual consumption in 1961-63 was 4,700 metric tons, the demand for pulp and paper is estimated by FAO at about 37,000 metric tons for 1975 and 75,000 for 1985. It is planned to construct a large export mill, based on mixed tropical hardwoods (capacity 170,000 tons per year) partly to meet the increasing demand for banana containers, but it is doubtful whether this project will materialise before 1975.

In Kenya the development of forestry and forest industries has been outside the coffee-growing areas. The requirements for coffee are somewhat different than for forest plantations, because softwoods are generally grown at higher altitudes and under less humid climatic conditions. In 1965 Kenya had a total of 70,800 hectares of exotic softwood plantations (*Cupressus lusitanica, Pinus patula* and *P. radiata*) planted mainly by the traditional taungya system by which shifting cultivation is carried out in combination with the raising of plantations of forest tree species. Government planting of exotic softwoods is continuing at the rate of approximately 4,850 hectares per year, with a planned increase to about 6,070 hectares a year by 1968 and an interim total target of 150,000 hectares by 1985. As a result, plantation logs are gradually replacing indigenous logs as the main source of raw material for the sawn timber industry. In 1965, out of a total recorded cut of 187,000 cubic meters, the contribution of plantation softwoods was 99,800 cubic meters. The sawmilling industry, originally designed to utilise large-diameter logs from indigenous forests, is gradually being redesigned for sawing softwoods. Three large private mills, utilising softwoods, are situated west of the Rift valley in the region of the biggest concentration of plantation timber. Main progress is likely to take place in the pulp and paper field. Imports of paper and paperboard now amount to some 37,000 long tons per annum, valued at £3.5 million.

The recent increase from the 1958 level of 17,000 long tons was stimulated by the establishment of local packaging manufacturing industries which consume over 10,000 tons of industrial paper and paperboard annually. A viable industry, with a mill capacity of 50,000 tons integrated with a sawmill with a capacity of 10,000 standards has been recommended by FAO for the district around Broderick Falls. In this connection, the FAO/IBRD Cooperative Programme has also suggested the establishment of some 30,000 hectares of softwood plantation. The proposed project represents part of the Kenya Government's long-term afforestation programme.

In Tanzania, the main Arabica coffee zone lies in the rain-shadow area of the Kilimanjaro mountain range. With regard to Robusta, the situation in Tanzania is similar to that in Kenya except that the indigenous forests still continue to be the principal source of industrial fuelwood. Softwood plantations, with the same species as in Kenya and under similar conditions of site and climate in the highlands, have been established in several places and they now total some 16,000 hectares, the current planting programme being 2,400 hectares per year. Wood-based industries are not strongly developed, sawmilling being the principal industry. Sawtimber exports in 1960-65 averaged approximately 21,000 cubic meters annually. No pulp or paperboard mills exist. The Government has under consideration an additional 20-year target of some 80,000 hectares of softwood plantation near the coast in the Ruvu region.

Possibilities of Diversification

The possibilities of diversification from coffee into forestry vary in their purpose and scope, depending on ecological and other conditions. In general, the Arabica coffee zones are equally suitable for forest crops, particularly for plantations of fast-growing species. The need for diversification into forestry is dictated by:

1. the operation of quotas under the International Coffee Agreement which calls for partial replacement of coffee by alternative forms of land use;
2. the necessity of replacing coffee grown under marginal conditions by a more profitable form of land use, which may require the stabilisation and restoration of suitable soil and the improvement of moisture conditions by the introduction of a forest cover.

The prospects of forestry have to be considered in the light of all land-use alternatives to coffee. Where there are expanding forest-based industries and exports, the case for forestry development is strong. However, with a few exceptions, forest industrial interests are generally outside the coffee zones.

The most notable exception is Brazil, where the main coffee-producing States, particularly São Paraná, are also the areas of greatest industrial development based on either natural forest resources or plantations. In all the coffee-growing zones in Brazil, the possibilities of diversification by afforestation have been recognised and included in the official IBC/GERCA Diversification Programmes, but so far without substantial results. These programmes also provide for appreciable financial support to wood-based industries, some of which are dependent on plantations.

In the States of Paraná, Minas Gerais and Espirito Santo, the accelerated expansion of afforestation is possible and desirable in order to replace eradicated or poor coffee. Such action should be based on existing experience and research concerning the choice of species and techniques. Substantial use of *Pinus caribaea* is indicated for expanded afforestation especially on degraded sandy soils, for example in the State of Paraná, In areas subject to intense heat in the summer and cold in the winter, afforestation is needed as a means of restoring the balance between forestry and animal husbandry, and also in certain localities for special reasons such as erosion control. Plantations are also necessary for the improvement of grazing conditions by providing shade in summer and shelter in winter.

As regards institutional needs, the Instituto Brasileiro do Cafe introduced in July 1967 a programme to guarantee interest-free financing for agricultural and forestry development in all areas covered by coffee diversification contracts. However, organisational measures to ensure that credit is made available would be even more important. Private forestry, which has been the backbone of forest industrial organisation in Brazil, appears to have the initiative, technical knowledge, and facilities for both external and internal financing to play a leading role in future development. However, close coordination in industrial planning is needed between the public and private forestry sectors in order to strengthen existing infrastructural and institutional arrangements.

In Colombia, plans for the expansion of industrial plantations in the main coffee-growing areas are essentially sound. In the principal coffee-producing Department of Caldas, the regional development programme makes provision for increasing the plantation area from 800 to 2,700 hectares by investing 13.2 million pesos. In the restoration of degraded pastures, *Alnus jorullensis* has been found to offer annual economic returns estimated at 800 to 1,000 pesos per hectares on a rotation of 25 to 30 years. Attention should be given to the possibility of expanding the present area of the indigenous *Guadua* bamboo by plantations, especially in connection with the rehabilitation of marginal coffee land or pasture. This bamboo wood should continue to meet increasing local needs and serve as raw material for small-scale rural industries such as low-cost housing and for the pulp and paper industry. Prospects for a planned expansion of plantations in the Department of Antiquoia are also good with the establishment of a new forestry corporation, it is planned to set up a large-scale newsprint plant based on long-fibered species.

In the Central American countries, the forestry aspect has so far been generally neglected in existing diversification projects, particularly in Guatemala. However, the main prospects for more diversified export-oriented development in Guatemala center on timber and wood products, especially in the undeveloped northern region, comprising hardwood resources outside the coffee zone. There are also unsurveyed pine resources closely connected with the coffee areas. Part of these resources have already been studied in connection with the private projects to set up a pulp and paper plant with the proposed production of 90,000 metric tons annually.

In Guatemala, and also in Mexico, which already has a well-developed forest industry, no incentives are provided for the establishment of plantations other than those needed for the restoration of eroded coffee lands. However, with the increasing use of the forest resources, chiefly softwoods, it is possible that private interests will be stimulated in planting pine species in the coffee zone. In El Salvador, with its very limited forest resources, a planned afforestation programme to augment the present pine resources is urgently needed. An estimated target of 42,000 hectares has been suggested.

In Africa, Arabica coffee is grown in Angola, Ethiopia, Kenya and Tanzania, but Angola and Tanzania are also Robusta producers. At present, the Arabica production in Angola is small, with some plantations in the central high plateau region. However, the ecological potential for growing coffee in this region is virtually unlimited. A coffee-forestry integrated approach is at present being adopted, and there is a spreading use of a good fast-growing timber species (*Grevellia robusta*) for shade. As a result, there is ample scope for a well-planned expansion of plantations of other fast-growing species, as against the slower growing cypress, to supply long-fibered requirements for the pulp and paper industry. Such a development, together with the planned expansion of the pulp and paper industry, current private investment plans for diversification into particle board and the creation of a sawmilling industry utilising eucalypts and cypress, could also help to utilise contract labor likely to be released from the Robusta areas.

In Ethiopia, the contribution of forestry to economic growth is likely to remain limited. The value of forestry output is estimated by FAO to rise annually from U.S.$34.2 million in 1962 to about $41.8 million in 1975 and $49.3 million in 1985. The export-oriented potential of Ethiopia's natural forests will remain small, although there are prospects of import substitution of wood-panel products by better organisation of supplies to the existing mill in Addis Ababa and the newly erected mill in Jimma. The planned construction of a paper mill in Wonji and the establishment of a few paper-converting plants may reduce the present complete dependence on imports.

Both in Kenya and Tanzania, forestry development is largely promoted and financed by the governments outside the coffee areas since the purchase of coffee land would be too expensive. National development plans already include the establishment of forest industries based mainly on softwood plantations in the same zone, often adjoining coffee areas.

In the Robusta zone of west Africa, forestry and coffee interests coincide in Ivory Coast. In this country, it is important to restore a balanced forestry development and to sustain, if not improve, its contribution to exports and the national product. For this purpose, it is necessary to curtail the expansion of agricultural crops, particularly coffee, which is overproduced and extended into marginal lands within productive forest areas. There is also a need for the improvement and rehabilitation of forestry resources. A scheme to establish about 58,000 hectares of industrial plantations is being financed by a newly formed state agency (SODEFOR) from a tax of 2 percent on the nominal value of exported logs. The project includes the planting of about 44,000 hectares of village pole and fuelwood plantations in the savanna zones.

However, the planned expansion of industrial plantations is mostly for teak, which has been grown mainly in the moist semideciduous forest as well as in the "derived" savanna zones. The enrichment of natural forests by suitable techniques, including the planting of fast-growing species, has not yet been attempted on a sufficiently large scale. Bamboo (*Bambusa vulgaris*) has been successfully introduced around Abidjan in small-scale plantations where it gives an annual yield of about 20 tons per hectare. However, there are no known plans for its extension for pulp and paper production.

16

Sustainable Coffee Marketing

The potential is enormous given the skyrocketing international growth in demand for speciality coffee. In the U.S., which accounts for about half the global market for roasted gourmet coffee, sales of such coffee increased from approximately $1 billion in 1990 to $2.5 billion in 1995. The question is how continued market expansion can be harnessed to promote forest conservation, environmental quality and higher incomes for coffee growers who implement methods to protect the environment.

Coffee drinkers who are concerned about issues such as migratory bird decline, pesticide impacts, or rural poverty in developing countries need to make their views known to the supermarkets and speciality coffee outlets where they shop. Consumers should press coffee retailers to provide specific information on the environmental and social conditions under which their coffee was grown and processed. Businesses and government agencies should consider the environmental dimensions when choosing the coffees they provide to their customers or employees. Shareholders in firms dealing in coffee should urge company managers to integrate environmental criteria into their commercial decisions.

The speciality coffee industry has yet to begin any large-scale, concerted initiative to promote environmental protection in coffee producing countries. Nevertheless, the Speciality Coffee Association of America recently established an Environmental Policy Task Force to address ecological issues associated with coffee. The International Coffee Organisation (ICO), the principal coffee trade group worldwide, held a seminar on "Coffee and the Environment" in May 1996 at the ICO's headquarters in London.

Environmentally conscious consumers should use the power of their purchasing decisions to support coffees that are certified organic, marketed through alternative trade or social justice channels, or backed by environmental criteria such as forest conservation or water pollution prevention. These three categories vary in areas of emphasis, but share a common thread of environmental protection.

Marketing Organic Coffee

At present, organic coffee accounts for just one or two percent of the $5 billion worldwide market for speciality coffee. However, organic coffee currently exhibits the fastest growth among gourmet coffee types. In addition to exports, significant growth potential exists in coffee producing countries where gourmet coffee is just beginning to emerge in domestic markets. For example, coffee co-ops from Chiapas, Mexico, have recently established retail outlets in Mexico City to sell their organically-grown "La Selva" coffee.

Production of certified organic coffee has expanded recently in northern Latin America. Such production can be found in all countries throughout the region. One example is Indígenas de la Sierra Madre (ISMAM), which is made up of 1,200 small-scale "campesino" coffee growers in Chiapas, Mexico. The ISMAM co-op exported 20,000 sacks of gourmet organic coffee in 1995 directly to Europe, the United States and Japan. Within Latin America, Peru outstrips other countries in terms of area, with nearly 44,000 hectares under certified production. Mexico, which produces nearly as much organic coffee as Peru, does so on just under 26,000 hectares. Other countries with land devoted to certified organic coffee include Guatemala, El Salvador (4900 hectares), Nicaragua (1400 hectares), and Costa Rica (550 hectares).

Certified organic coffee fetches significant price premiums on the order of 10 to 15 percent above gourmet coffee without the organic trademark. The price premium often translates into substantially higher returns for coffee growers, although the net benefits of moving to certified organic production can vary substantially from producer to producer, depending on added production costs and other variables. Organic coffee co-ops pay thousands of dollars each year to cover certification costs such as the time and travel expenses of field inspectors. The downside to organic certification from many growers' perspective is the cost of periodic inspection. For the multitude of small coffee growers who are *de facto* or "passively" organic producers because they cannot afford to use agrochemicals, inspection costs can present a formidable obstacle to certification, and hence to the premium price they might otherwise obtain for their coffee.

Organic coffee growers are typically organised into local cooperatives that are affiliated with, and bound by the standards of, international certification programs. The largest of such programs is Organic Crop Improvement Association International (OCIA), which as of late 1995 claimed more than one million certified hectares (2.5 million acres) and 30,000 grower-members worldwide. Other programs certifying organic coffee in northern Latin America include the European-based Naturland and Demeter.

The international certification programs serve several functions. For example, OCIA sponsors crop improvement seminars and other technical assistance for farmers implementing organic systems; independent third-party inspection of certified farms, with an audit trail to track coffee and other commodities from consumers to producers; and a trademark that appears as a label on OCIA-certified coffee and other organic products.

Sound environmental stewardship is a central tenet of the organic agriculture movement. For example, soil building practices are key OCIA requirements for certified organic farms. The OCIA standards permit certification only of fields or farms where no synthetic pesticides or fertilisers have been applied during the preceding three years. Diversified forest cover appears to be one common approach by which organic coffee farmers achieve certification standards for soil quality and chemical use. For example, organic coffee growers in Mexico maintain diverse shade cover to enhance soil fertility and to reduce their production systems' vulnerability to pests. OCIA currently encourages its coffee producing members to diversify the shade cover, so that growers can benefit from a variety of products associated with their holdings. However, existing organic standards do not contain explicit, measurable criteria for diversified shade cover.

The soil building techniques used in organic coffee farms often help reduce the waste stream of pollutants to water supplies. The Asociación de Caficultores Orgánicos de Colombia, a co-op in the process of obtaining OCIA certification, composts all organic waste from coffee processing to create a rich mulch for use as a natural source of nutrients for coffee plants. Nevertheless, measurable pollution prevention standards are not part of existing organic certification regimes.

Alternative Trade and Social Justice Market

Coffee producers in certain countries enjoy premium prices for their coffee due to the connections they have forged during the last decade with groups that make up what is known variously as the "solidarity," "social justice," "alternative trade," or "fair trade" movement. The movement is based on the idea that producers of traded commodities in developing countries are capable of achieving economic success provided they receive fair prices in international markets for what they produce. Recent years have seen a growth of the movement, with trade unions, church groups, and women's organisations becoming involved.

Throughout Europe, for example, fair trade coffee accounts for 11,000 metric tons of traded coffee annually, finding outlets in some 35,000 supermarkets. Like coffee that is certified organic, coffee distributed through alternative trade channels currently represents a very small fraction of the worldwide speciality coffee market. Sales have increased, however, as more and more coffee drinkers have learned about the poverty and dismal working conditions characterising small coffee producers' lives. Global sales of coffee in the social justice market amounted to $400 million in 1995, according to estimates from the International Federation of Alternative Trade (IFAT), an association that oversees 36 alternative trade organisations worldwide through a code of ethics established in 1990 and updated in 1995.

The social justice market is organised around the International Coffee Register, which is essentially a company owned by the fair trade groups Max Havelaar, TransFair, and the Fair Trade Foundation. A total of 286 coffee-producing cooperatives are members of the Register, representing about half a million growers around the world. Under current arrangements, grower groups are guaranteed $1.26 per pound for "green" (ready-to-roast) coffee. If world prices average above this figure, producers receive five cents per pound above the world price.

There are currently about 15 licensed importers of fair trade coffee. If a producer cooperative needs a cash advance to use for purposes of extending credit to individual growers, or for other expenditures, it falls to the importers to provide advanced funding that can total up to 60 percent of the contracted coffee with that cooperative, at rates of interest negotiated between the importer and the coffee cooperative.

The social justice movement and its corresponding market are much more developed in Europe, where they originated, than they are in the United States and Canada. European solidarity groups have long been active in Latin America, with institutional support available to them through social democrat governments and a solid base of non-governmental organisations. Community development projects funded by the private sector or via government funds have long been a hallmark of development work conducted by countries like Belgium, the Netherlands, Germany, France and Great Britain.

In 1988, Max Havelaar Netherlands developed a market label to link small coffee growers in Mexico directly to international markets. The label propelled the Dutch "alternative" trade in coffee from a mere 0.3 percent of national consumption to 2.3 percent by 1995. In Switzerland, it has captured 5 percent of the market. The Max Havelaar label now has a presence in at least six European countries, including its home country, Belgium, Austria, Germany, France and Switzerland. Another social justice label, TransFair, exists in eight nations around the globe, including Canada, Japan, the US, Italy, Germany and Austria.

An innovative direct marketing strategy has been employed by Aztec Harvest Coffee Company, which is owned by Mexican small-farmer cooperatives and is structured to bypass intermediaries in selling coffee to US and European buyers. Substantial portions of the coffee traded through Aztec Harvest come from certified organic co-ops. Well-known Aztec Harvest customers have included Ben & Jerry's and United Airlines.

Although focused primarily on social justice for small coffee growers, the alternative trade movement has incorporated environmental objectives in a general way. For example, criteria for the International Fair Trade Coffee Producers' Register specifies that any producer organisation wanting to participate must be committed to "sustainable development strategies, applying production techniques which respect specific ecosystems and contribute to the conservation and a sustainable use of natural resources, in order to avoid as much as possible — or even totally — the use of chemical inputs."

Moreover, the IFAT's Code of Ethics spells out environmental expectations for the participation of alternative trading organisations (ATOs). The two-point environmental section of the code states:

a) It is also the aim of ATOs to encourage the production of goods by means which preserve the environment and conserve scarce resources and in ways which cherish the skills and develop the capacities of the producers and do not harm their health. This applies equally in the First World as in the Third World.

b) ATOs are committed to encouraging development which is sustainable and responsible in terms of the long term survival of the human species and of the natural world.

In many cases, strong overlap exists between the certified organic and social justice coffee movements. The Massachusetts-based Equal Exchange works primarily to ensure fair prices to small growers, and deals mostly in certified organic coffees. In other cases, the fair trade market makes linkages with small peasant producers who, because of their inability or unwillingless to use costly chemical inputs, produce what can be regarded as a "passively" organic or "organic by default" coffee.

Environmental Criteria in Consumers' Coffee Choices

There are many good reasons to buy certified organic or social justice coffee. There is also a need to develop and apply a broader range of environmental criteria than may be covered by existing certification regimes. For example, explicit criteria for coffee produced through "shade-grown" or "bird friendly®" management systems could provide a powerful market force for forest conservation and sustainable economic development in northern Latin America and other coffee producing regions. Additional criteria are needed to reflect whether pollution prevention measures have been applied in coffee processing.

To ensure consumer confidence, any new environmental criteria for coffee must be measurable, scientifically rigorous and consistent. The criteria should not be one-dimensional, but rather should reflect the complexity of coffee management systems and the biodiversity values of various levels and composition of shade cover. Additionally, the criteria should be evolutionary, that is, subject to refinement in the face of improved scientific understanding or technological innovations.

Environmental criteria might find expression in the marketplace in a number of ways. Adding or overlaying new environmental metrics within existing certified organic or alternative trade systems is one possibility. Another is a separate certification regime such as the Rainforest Alliance's "ECO-O.K." Program, which is moving toward certifying coffee with criteria developed by Fundaci-n Interamericana para Investigaciones Tropicales, a Guatemalan NGO. Draft criteria for the "ECO-O.K." effort on coffee include maintenance of a minimum number of shade trees per hectare and encouragement to growers to minimise agrochemical applications. The use of native perennials as shade, as well as the maintenance of vegetational buffer zones next to rivers, streams, and lakes constitute other criteria stipulated for "ECO-O.K." certification.

One option worth serious consideration is a system that, as distinct from dichotomous certification regimes (where products are either certified or not), rates coffees according to their performance on a range of environmental parameters. Hypothetically, for example, a scoring method might be devised where a coffee receives "one star" if grown under shade cover with limited diversity, or "two stars" if the shade cover features a highly diverse plant community and forest canopy structure. Similar gradations could be developed to reflect varying degrees of pollution prevention and waste recycling in coffee processing.

These options are not necessarily mutually exclusive. However, any further integration of environmental criteria into coffee markets must proceed in a way that informs rather than confuses consumers. Moreover, such integration must ultimately satisfy coffee roasters' primary interest in product quality and competitive prices. It would be unrealistic to assume that new environmental criteria, no matter how compelling, will be sufficient to "pull" particular coffees through the market irrespective of price and quality considerations.

17

International Institutions

Governments and international institutions mobilise their efforts to support traditional coffee farmers whose agricultural practices preserve biodiversity and enhance soil quality in all over the world. Individuals and institutions should likewise direct their efforts into boosting methods of coffee processing that substantially reduce solid and liquid waste dumping into water supplies. Local governments, bilateral funders, and multilateral institutions should find ample reason to target environmentally sensitive coffee production and processing as acceptable funding opportunities.

Establishing policies that preserve shade coffee production and allow small coffee growers to benefit from the good land stewardship practices that many already have in operation not only will help maintain the biodiversity associated with shade coffee, but will bring greater economic benefits to the coffee communities involved. Moreover, ecologically sound shade coffee production falls squarely within the realm of carbon sequestration, and producers could potentially fit within programs such as the United States Joint Implementation Initiative.

A priority will be to provide financial incentives, credit, and technical assistance for coffee growers and processors to adopt methods that maintain crop yields and profits while benefiting environmental values. Agriculture, tax and trade policies should be reformed to encourage, rather than create barriers to, environmentally sensitive coffee production. Support is also needed for expanded research on the ecology and economics of coffee production and markets in the Americas.

Funding Incentives for Coffee Production

Adoption of sustainable coffee systems presents transitional economic risks for growers and processors. Targeted incentives are needed to enable more small-scale coffee growers in Latin America to maintain or implement environmentally sound management practices that help sustain long-term productivity and profitability. Sustainable development in the Latin American coffee sector should be a priority for bilateral economic assistance, the Global Environment Facility, multilateral development banks, and national environmental funds.

1. Bilateral Assistance

Through its various programs for bilateral foreign assistance, the United States should support sustainable coffee production in Latin America. Recent examples from the Inter-American Foundation and the U.S. Agency for International Development illustrate the potential for bilateral funding to improve economic and environmental conditions for coffee producers. The need for incentives connected to environmentally sensitive coffee argues strongly for increased bilateral assistance for sustainable development in developing countries. Unfortunately, current political trends in the US are toward sharp reductions in foreign assistance funding.

The role of NGOs operating internationally will become increasingly important in providing financial and technical assistance for environmentally sound coffee production. For instance, Conservation International is developing a multi-faceted program to work with local farmer organisations in priority biodiversity areas in Latin America and worldwide. The program will be designed to provide pre-harvest financing to enable growers to store their coffee until they can obtain a favourable price from buyers, technical support in environmentally sound production methods, and transitional costs for implementing positive land stewardship practices. In Mexico, the Coordinadora de Peque-os Productores del Café de Chiapas (COOPCAFE) is undertaking a program under a grant from the John D. and Catherine T. MacArthur Foundation to train peasant farmers in that state in organic coffee production and to support ecologically sound agriculture by farmers operating in the buffer zone of the Lacandon rain forest.

2. Global Environment Facility

The Global Environment Facility (GEF) is the world's only multilateral dedicated fund for the environment, providing grants to developing countries to protect global resources. The GEF is governed by an independent Council, and the projects it funds are implemented by the World Bank, United Nations Development Program (UNDP) and United Nations Environment Program (UNEP).

The GEF was launched in 1991 as a pilot program. In 1994, the Facility completed a restructuring process that reflected many recommendations by conservation groups and governments, as well as an independent evaluation of the pilot phase. Total funding for the GEF over the next three years is set at $2 billion — an amount to be contributed in annual installments by 26 nations. The U.S.-agreed share of the GEF replenishment is $430 million, of which Congress appropriated $90 million for fiscal year 1995, but just $35 million for fiscal year 1996.

The GEF's purpose is to address environmental problems that are "global" in nature. Hence, GEF-funded projects must fit within one or more of the Facility's focal areas of biological diversity, climate change, international waters, ozone depletion, and land degradation. Environmentally sound coffee production should fall squarely within the GEF guidelines, in the biodiversity area and perhaps others. In its discussion of forest ecosystems, the operational strategy recently approved by the GEF states that priority will be given to "conservation of areas of importance for migratory species." The strategy's list of activities for sustainable use of biodiversity includes "[P]romotion of sustainable production and use of natural products, such as nontimber forest products. . .and agrobiodiversity-related products, including the development and implementation of sustainable harvesting and marketing regimes." These and other passages in the document are directly relevant to coffee production in northern Latin America.

Global Environment Facility grants are available not only to national governments, but also to non-governmental organisations and private sector businesses. A small fraction of GEF funds are allocated through a UNDP-administered Small Grants Program for projects conducted by non-governmental organisations and local communities. Conservation groups have urged enlargement of the existing Small Grants Program, as well as establishment of a Medium Grants Program for NGO projects whose scale falls between the small grants program and the regular GEF funding available to governments. The small and medium grants approach would be well-suited to funding local environmentally sensitive coffee enterprises in Latin America.

Another potential funding window for ecologically sound coffee is a newly proposed "Biodiversity Enterprise Fund for Latin America," to be directed by the International Finance Corporation (IFC) with partial support from the Global Environment Facility. Affiliated with the World Bank, the IFC is the largest multilateral source of private sector financing in developing countries.

Preliminary documents describing the Biodiversity Enterprise Fund for Latin America suggest there may be significant potential for financing of coffee operations that protect ecological values. The project proposal submitted to the GEF notes specifically that "fast growing markets for biodiversity-linked products" are creating new opportunities for projects that advance both conservation and development objectives. The IFC's draft feasibility study for the project states that, among its priorities, the fund will support investments in "alternative, certified organic, or biologically diverse agricultural methods."

3. Multilateral Development Banks

Within their Western Hemisphere lending, the World Bank and Inter-American Development Bank (IDB) appear to be increasing their programmatic emphasis on biodiversity and other environmental objectives. The 1987-1993 World Bank portfolio contains numerous projects involving agriculture and forest management in Latin America and the Caribbean. The IDB's draft plan for implementing the Action Plan from the 1994 Summit of the Americas states that the Bank will "make biodiversity concerns an integral element of rural development initiatives."

Support for sustainable coffee production should be a priority for multilateral development bank programs in Latin America. World Bank and IDB loans for forestry and agriculture projects in the region should be assessed for potential impacts on small-scale, traditional coffee growers and local communities. The Banks could play a particularly important role in addressing funding needs for projects requiring significant capital investment, such as pollution prevention technology in coffee processing facilities. For example, water use reduction and recycling are part of an IDB-funded project being implemented by small-scale grower cooperatives in El Salvador in partnership with Appropriate Technology International.

The World Bank is currently developing a $200 million Agriculture Development and Rural Poverty project for Mexico. Slated for approval in mid-1996, the project targets the extremely poor rural areas in Mexico's southern states that encompass much of the country's coffee-growing lands. If properly designed and implemented, the project could benefit sustainable, shade-grown coffee enterprises. The Bank's public information document describing the project observes that "government price and trade policy for agriculture has favoured larger commercial farmers. . .as well as commercial livestock producers" and, moreover, that "[D]uring the past six years, government agricultural programs aimed at poor producers have diminished in scope and in effectiveness."

The project's defined objective is to promote long-term sustainability for poor, small-scale farmers and rural organisations through measures such as soil and water conservation, and improved processing and marketing of agricultural commodities. Examples of possible areas for project financing include on-farm investments in sustainable natural resource management, applied research and agricultural extension, and assistance to producer organisations to improve their capacity for marketing, access to credit and adoption of sustainable production technologies.

Two additional funding institutions, created under the environmental agreements accompanying the North American Free Trade Agreement (NAFTA) among Mexico, Canada, and the United States, have potential to fund sustainable coffee production in Mexico. One is the North American Development Bank (NADBank), whose estimated $7-8 billion in total financing capability over the next decade will be directed toward funding public works projects for sewage treatment, solid waste management and other environmental infrastructure projects needed in the U.S.-Mexico border region. The Bank nevertheless features a "Community Adjustment and Investment Program" under which 10 percent of the NADBank capital is supposed to go toward grants and loans to communities throughout Mexico and the U.S. affected by economic integration. Support for communities in southern Mexico engaged in environmentally sound coffee production should receive priority attention within implementation of the NADBank's community adjustment window.

The other institution of concern is the Commission for Environmental Cooperation (CEC). At the October 1995 meeting of the CEC Council, the environmental ministers from the three countries jointly announced creation of a $2 million North American Environment Fund. Grants from the fund will support local community organisations "for projects that promote an integrated approach to ecosystem management and the sustainable use of natural resources important to the region as a whole." Use of the fund to support sustainable, shade-grown coffee would fit well with the Commission's current priority focus on protection of migratory bird habitat.

4. National Environmental Funds

An additional funding source for ecologically sensitive coffee production may be found in national environmental funds, which comprise various mechanisms such as trust funds, foundations and endowments that provide financial support for activities that benefit biodiversity conservation and other environmental purposes. Such funds are typically governed by Boards of Directors with governmental and non-governmental representatives, are able to receive and manage money from a variety of sources, and can disburse grants to non-governmental organisations and community groups.

National environmental funds have demonstrated considerable promise in financing local conservation and sustainable development projects, building NGO capacity, and strengthening democracy and the role of civil society in conservation policy choices. The role of national environmental funds becomes even more compelling in the face of cutbacks in U.S. bilateral and multilateral assistance for sustainable development in developing countries.

Recent years have witnessed a rapid and diverse proliferation of national environmental funds in Latin America. One example is ECOFONDO, which is a private trust fund established in 1993 and managed jointly by the Colombian government and the NGO community. Major funding sources for ECOFONDO have included forgiveness of official USAID debt through the Enterprise for the Americas Initiative, and a subsequent grant from the government of Canada. Environmentally sound coffee production certainly falls within several categories of activities eligible for ECOFONDO grants, such as "sustainable development of watersheds," and "conservation and sustainable management of ecologically important areas."

Reforming Agriculture and Tax Policies

It is essential that national farm policies — through whatever mix of price and income support, commodity supply management, credit assistance, research, extension, conservation cost-sharing or other instruments they may employ — be structured to encourage rather than penalise traditional coffee producers who practice good land stewardship. Government intervention, for example, should be targeted to boost the productivity of under-managed, "passively" organic coffee farms that, for reasons of poverty and/or neglect, have not attained adequate yields. Another priority is to ensure that small-scale coffee growers have access to adequate credit to enable them to implement environmentally sensitive practices and obtain good prices for their coffee.

Unfortunately, agricultural policy frameworks in Latin America have done little to provide incentives for sustainable coffee producers. It has instead been more common for commodity subsidies and related programs to reward large-scale operations and create bias against small-scale, diversified farming systems.

Mexico is a case in point. Until the early 1990s, Mexico subsidised livestock heavily through artificially low feed prices; this created implicit taxation of shade coffee crops and encouraged conversion of forests to pasture. Notwithstanding benefits delivered to some small producers, the National Solidarity Program implemented under the Salinas administration has been criticised for not addressing underlying problems facing the Mexican coffee sector such as the untenable debt loads of many small growers. Recent social unrest in southern Mexican states such as Chiapas and Guererro, has been attributable, at least in part, to the long-standing failure of government policies to address the extreme poverty among coffee growers.

During the 1970s and 1980s, the Mexican government's Instituto Mexicano del Café (INMECAFE) promoted intensified coffee production through the reduction or removal of diverse shade cover, planting of high-yielding hybrid coffee varieties, and increased agro-chemical inputs. Privatisation of INMECAFE went forward in the late 1980s and early 1990s without effective safeguards against small farmer dislocation. In a move that seems reminiscent of the production-oriented policies of the 1970s and 1980s, the Mexican government recently announced an ambitious proposal to provide coffee plants, credit and technical assistance to growers wishing to renovate their operations in search of higher yields. As part of a larger production-oriented program known as "Alianza Para El Campo", planners believe this new policy can position Mexico's total coffee output ahead of Colombia's within the next decade. Experts have criticised this initiative, among other reasons, for ignoring its negative environmental implications.

Since 1993, Mexico has been instituting a package of agricultural reforms known as PROCAMPO (Programa de Apoyos Directos al Campo, or "Program of Director Support to the Countryside"), designed to replace commodity price supports with fixed income support payments made directly to farmers over the next 10 years, and phased out over the subsequent five years. In one change that could conceivably benefit the competitive position of organic coffee growers, the PROCAMPO reforms remove most subsidies for inputs such as purchased chemical pesticides.

The overall impact of PROCAMPO on the environmental practices of coffee producers is not immediately clear because the program is so new and its early implementation has been complicated by the Mexican economic crisis following the December 1994 peso devaluation. Changes announced by a Mexican inter-governmental commission in October 1995 will broaden the scope of PROCAMPO to include greater emphasis on technical assistance, decentralisation of decision-making authority to the state and local levels, and a new program called "PRODUCE" under which the government will pay up to 50 percent of certain costs incurred by the poorest farmers.

Because its benefits are not linked to production of specific commodities, PROCAMPO is expected to reduce incentives for surplus production of major crops such as corn, soybeans and sorghum. This represents a shift away from certain previous policies that, as noted above, have favoured large-scale grain producers and commercial livestock operations at the expense of forest conservation and diversified farming systems including coffee. On the other hand, some non-governmental organisations and community forestry associations in Mexico have expressed concern that PROCAMPO encourages conversion of land to production of agricultural commodities as opposed to sustainable forestry enterprises.

Reforms to change the incentive structure for coffee producers will not be complete without a review of tax policies. This examination should focus on removing perverse tax preferences that favour unsustainable farming methods, as well as on providing tax benefits to encourage development of sustainable land use and resource management.

One promising strategy would be the creation of sustainable agriculture funds through taxes on pesticides. Such revenues could be channelled into national environmental funds, as mentioned above. This would be a means of "internalising" the costs to environmental quality and human health that result from chemical-intensive sun coffee production. Useful precedents for taxing agricultural chemicals exist in several European countries, as well as in states such as California and Iowa.

A related approach is to recapture a portion of revenues associated with international coffee trade, to be directed toward social and environmental purposes. For example, Colombia has used savings from the 1990 elimination of the European Community's four percent tariff on coffee imports to establish an Ecological Fund for Coffee Zones. Interest from the investments of these revenues now supports projects focusing on the integrated management of watersheds, the recycling, composting, and management of municipal garbage, water treatment, and community development via environmental education programs.

Finally, agricultural and tax policy reforms should be accompanied by measures to strengthen regulation and monitoring of pesticides. Tighter restrictions are needed on exports of banned pesticides, in addition to upgrading of domestic regulatory regimes. Another priority for countries throughout the Americas is to implement "right-to-know" requirements that assure provision of reliable pesticide use information to farmworkers, regulators and the public.

New Research Strategies

Scientific evidence now available makes a compelling case documenting the environmental damages of industrial coffee production, and the corresponding environmental advantages of traditional methods. The various studies conducted to date provide a strong basis to move forward, without delay, to mobilise market and policy forces on behalf of sustainable coffee systems.

A coffee strategy to promote environmental protection and sustainable development in northern Latin America, however, will require a commitment to expanded research. This commitment is needed to improve our understanding of the complex agro-ecology and economics of coffee, and thus to give needed direction to choices made in policy, funding and markets. The research strategy should be geared toward providing practical information to coffee growers on how they can implement environmentally sound production systems, and achieve success in domestic and international markets.

The following are examples of priorities for additional research on environmentally sensitive coffee:

- — mapping of Latin American countries to determine the current spatial distribution of traditional, shaded, biodiverse coffee lands;
- — research on the tree species known to be associated with greater biodiversity of birds, insects, etc.;
- — studies to determine the appropriate levels of thinning or pruning of the overstory trees that will maximise associated biodiversity and coffee production;
- — the role of physiological characteristics of shade species, such as flowering and fruiting patterns, to

determine what mix of shade best enhances biodiversity levels;

— studies comparing the performance of shade and sun coffee on environmental parameters such as migratory species, biological diversity, soil productivity and water pollution;

— market analysis for environmentally sound coffee and the non-coffee products raised on coffee farms;

— economic studies to measure the external costs of coffee production, and to describe options for internalising those costs; and

— studies comparing the economics of environmentally sound versus conventional production systems at the farm level.

Research strategies in these and related areas will necessarily include documentation of traditional knowledge by working with local people and communities; development of incentives for case studies on working farms and farmer-to-farmer information exchanges; increased support for research and monitoring activities over long time horizons; and establishment of regional networks of researchers and data bases for biogeographical and market information.

International Coffee Markets

International coffee markets are subject to wide fluctuations, in general adhering to supply and demand relationships. Major swings in coffee prices may go largely unnoticed by consumers in wealthy industrial countries, but are critical to individual coffee growers and national economies throughout much of Latin America. When coffee prices drop precipitously, the impacts fall hardest on low-income farmers and labourers. The consequences are frequently acute for small-scale growers practicing environmentally sound coffee production.

Since the early 1960s, a series of international agreements have failed to achieve long-term solutions to price instability in the coffee sector. Unemployment and financial impoverishment resulting from the collapse of the International Coffee Agreement in 1989 affected millions of farmers and farm workers in the coffee growing countries of Latin America. Prices have risen dramatically since the 1993 Coffee Retention Plan, which is an accord among 28 coffee producing countries from Latin America, Asia and Africa that seeks to restrict international market supplies by requiring members to withhold 20 percent of coffee stocks from exportation. However, much of the recent boost in prices is attributable to a mid-1994 freeze that destroyed a large portion of the Brazilian coffee crop. It is too early to assess whether the 1993 scheme will lead to more enduring coffee price stabilisation than the previous international agreements.

The world's trading partners have taken initial steps toward integrating environmental concerns into the International Coffee Agreement. Article 35 of the 1994 version of the Agreement requires members to:

> [g]ive due consideration to the sustainable management of coffee resources and processing, bearing in mind the principles and objectives on sustainable development agreed at the Eighth Session of the United Nations Conference on Trade and Development and the United Nations Conference on Environment and Development.

Agenda 21, the comprehensive action plan adopted at the 1992 United Nations Conference on Environment and Development in Rio de Janeiro (the "Earth Summit"), contains abundant provisions that apply to ecologically sound coffee. Examples include commitments aimed at "international cooperation to accelerate sustainable development in developing countries," "combating deforestation," and "meeting agricultural needs without destroying the land."

Future coffee trade negotiations should better serve the interests of sustainable coffee producers, with the goal of ensuring prices that enable a reasonable return on sales and that reflect costs incurred in making production systems environmentally sound. Toward this end, for example, international agreements should never undercut the ability of producers to bypass the "middle man" in international commodity markets, and to sell their products directly to roasters and consumers wishing to purchase coffee that is grown and processed without environmental degradation.

One option for future reform of coffee trade regimes is to promote agricultural shifts from sun-grown coffee into environmentally sound alternative crops. In addition to reversing the environmental damages from industrial coffee plantations, such shifts could help reduce surplus coffee volumes and thus contribute to more lasting price stability for sustainable coffee producers.

18

Economic Landscape of Coffee

Today, coffee still forms the economic backbone of many countries. The "traditional" coffee of today has actually evolved since this African shrub was first brought to the New World in the 18th century. Early on, coffee was often placed in the open sun, a practice that growers soon changed by planting shade or placing coffee in areas only partially cleared of their native forests. By the dawn of this century, a production system characterised by diverse shade and biological richness dominated the coffee landscape.

But by the 1970s, things began to change quickly and dramatically in Northern Latin America. Coffee farmers, just like producers of other crops around the world, found themselves face to face with the forces of modernisation. Added emphasis to change production technology came from agronomic problems like threats from diseases.

The result has been a sweeping tendency to shift from traditional production techniques to more modern ones, a move that usually involves changes in the shade cover, its management, and the use of agrochemicals.

Nature of the Traditional Coffee Farm

Traditionally, the structural profile of a coffee farm in northern Latin America has resembled that of a forest. With coffee as the understory shrub, a mixed shade cover of fruit trees, banana plants, and towering hardwood species forms a forest-like agroecosystem. Such an agroforestry structure results in a fairly stable production system, providing protection from soil erosion, favourable local temperature and humidity regimes, constant replenishment of the soil organic matter via leaf litter production, and home to an array of beneficial insects that can act to control potential economic pests without the use of toxic chemicals. Traditional coffee, in fact, has been cited as the region's most environmentally benign and ecologically stable agroecosystem.

A strong consideration for the small grower is what the shade coffee system produces in addition to the coffee harvest each year. Indeed, several "non-coffee" products are harvested on a continual basis from traditional coffee holdings. This diversification helps shield small producers from risks arising from the vagaries of nature, international market fluctuations, or societal structures. For many coffee producers in Central America, the mixed nature of shade cover traditionally maintained in coffee provides insurance against uncertainty, and maximum use of limited land holdings becomes an effective survival strategy. The coffee harvest provides income each year, the absolute amount of which depends upon yields and international prices. Other plants and trees in the coffee holding provide a host of products that the grower would otherwise have to buy on the local market. A farmer's entire family is often involved in traditional coffee production, especially for the harvest.

Income from selective timber harvest derived from shade trees can be substantial. In studies based on Costa Rican practices, timber stands of *Cordia alliodora* used as shade in densities of 120-290 trees per hectare, can produce a sustainable output of 6-15 cubic meters per hectare per year of commercial wood. Timber output such as this can help provide income security for small farmers; for instance, timber harvests from shaded cacao plantations saw Costa Rican producers through several years of tough financial times in the early 1980s, when plant disease decimated cacao production.

The species composition and structure of a traditional coffee system will vary according to country, ecological zone and grower. But examples from several places within northern Latin America point to the similarities of traditional coffee regardless of location. In particular, producers in many countries make use of shade trees in their coffee holdings, with smaller producers tending to make use of a variety of trees that provide edible fruits. In Nicaragua's southern uplands known as the Carazo district, just south of Managua, traditional coffee holdings have at least 25 species of fruit and timber trees associated with them, many of which are native species to this seasonally dry forest zone.

Throughout the region, many farmers plant nitrogen-fixing shade trees belonging to genera such as *Inga, Gliricidia, and Erythrina*. Small growers prefer to have fruit trees as well, such as citrus, bananas, and guavas. A study in Venezuela showed growers choosing a mixture of shade trees (distinct from fruit or timber species), fruit trees, bananas and timber species, regardless of the ecological zone in which they happened to produce. The density of shade trees approached 353 per hectare, and total tree density in this study reached 561 per hectare in some farm systems.

Features of Technified Coffee

Beginning in the mid-1970s, a successful push to "renovate," "technify" or "modernise" the coffee sector in much of northern Latin America emerged. The force behind this move came from a fungal disease known as coffee leaf rust (*Hemileia vastatrix*), also known by its Spanish name, *la roya* ("the rust"), spreading throughout the region. The devastating potential of the rust was known from historical records in India and Sri Lanka, where the disease halted coffee production within two decades in the second half of the 19th century. The wind-borne spores finally made a New World landfall in 1970 on the east coast of Brazil, an event that triggered panic throughout the coffee industry in the Americas.

Outbreaks and spread in Nicaragua's southern coffee district of Carazo in 1976 heralded the arrival of coffee leaf rust in Central America, prompting an urgent search for solutions. The most popular response was a technological one proposed by the United States Agency for International Development. With the coordination and financing of USAID or on their own, governments throughout the region implemented or participated in programs to technify their coffee. For most of the region, coffee leaf rust has not posed the problems originally anticipated. This most likely is due to the high elevation and/or the prolonged, intense dry season — physical conditions not conducive to the disease's proliferation — characterising much of the coffee zone.

Technification — now more commonly referred to as "modernisation" — consists of the replacement of old, traditional varieties of coffee such as *típica* or *bourbón* with newer varieties that respond well to chemical fertilisers. Another feature of the modernisation process involves the elimination or reduction of shade, the goal being to open the coffee up to the sun to deter the spread of fungal diseases, and to increase coffee yields.

In many cases, shade removal and establishment of high-yielding coffee plants have not achieved the intended objectives. For example, technification has often entailed planting of *caturra,* a dwarf mutant coffee variety discovered in Brazil during the last century and brought to Central America in the 1950s, which yields about 30 percent more coffee per shrub if supplied with fertiliser inputs. Although initially touted as being resistant to coffee leaf rust, *caturra* is, in fact, susceptible to the disease.

The transformation resembles in many ways the changes that began in basic grain production throughout much of the developing world in the 1950s. The high-yielding varieties of coffee, the use of agrochemicals, and the restructuring of the production unit itself all have their parallels in the "green revolution" associated with corn, wheat and rice production in the South. For coffee, the transformation means increasing the density of coffee plants from 1100-1500 per hectare to 4000-7000 plants per hectare. These higher-yielding varieties are planted very close together and typically plied with petroleum-based fertilisers, as well as herbicides, insecticides and fungicides. As discussed in the following section, these chemical inputs can create their own challenges, not the least of which is toxic exposure for farmworkers due to lack of information about the use and effects of chemical products or the infeasibility of wearing protective clothing in hot weather.

The transformed coffee landscape elicits images of industrial agriculture. The neat rows of coffee beneath direct sun or scant shade resemble an English hedge-row compared to the shrubby understory of a traditional farm.

Shade trees, when present in this industrial system, receive "scientific" pruning techniques that produce a thin laminar look to the canopy, thus reducing the structural diversity of what might otherwise offer an array of niches to insects, birds and other animals. Moreover, the limited shade trees sometimes retained in technified sun plantations often tend toward a single species. As discussed below, the modern coffee agroecosystem features much lower levels of structural and species diversity than the traditional coffee farm.

From Forest to Open Field

Conditions in the real world are more complicated than a strict dichotomy of "traditional" versus "modern" or "shade" versus "sun" coffee. Many examples of both these management systems exist across the coffee landscape of Latin America. In practice, however, they represent the extreme ends of a continuum of intensification within the coffee sector, and, as with any continuum, plenty of examples fall between the two extremes. Many issues will influence the final appearance of the coffee holding, including conditions associated with geographical setting — that is, factors such as topography, ecological zone, and rainfall.

The collective knowledge of a region's coffee growers, and the social and political interactions among the producers, will also shape a particular area's general management style. Likewise, institutions can affect where along the continuum coffee holdings in a particular zone may fit. Of course, an individual grower's own assessment of how to manage a coffee farm plays a central role in the final outcome of its appearance. This often involves inter-generational communication of knowledge. Fieldwork reveals that many growers manage their holdings in a way taught them by their fathers, uncles, or grand-fathers. Whatever the various factors, the reality of the coffee landscape across much of Latin America is one of diverse management styles.

One categorisation of these management styles that serves as a starting point for understanding this "management spectrum" was devised years ago by some Mexican coffee researchers and technicians. It identifies five different management types, using shade levels and management as indicators for intensification. The less shade there is, the more intensified the production system. Implied in the scheme is a tendency to be more dependent upon the market, and less inclined to produce a variety of commodities for household use, as a holding is more intensified or industrialised.

"Rustic coffee" displays the least intensified management system, which is characterised by the coffee plants being inserted into the existing forest with little or no alteration of the native, already-present vegetation. Production under these conditions is destined for the market, but little time and less capital is invested in realising this production. The "traditional polyculture" holding mimics the rustic coffee in structure, but the species diversity can be much greater because of the deliberate planting of other plants valuable to the household.

Yields in "commercial polyculture" systems are usually higher than those found in the less-intensified holdings, but commercial polycultures also include several non-coffee products that provide food and/or income for the grower. A "reduced" or "specialised shade" system normally displays a single canopy species (e.g., genera such as *Inga, Erythrina, Gliricidia, Grevillea*), the maintenance of which is highly controlled, giving an overall manicured appearance. This distinctive system often has a laminar look to the shade layer, and is in effect a two- or three-species agricultural system. Finally, there is the "open-sun" management practice, which eliminates the overstory completely. This system resembles tightly packed hedgerows, is highly productive if given the requisite chemical inputs, and is, like the reduced shade system, oriented solely to producing coffee for the market.

Impact of Institutions

The transformation of traditional coffee into a more intensive production system has a host of proponents. In Mexico, the now-defunct Instituto Mexicano del Café (INMECAFE) advocated the adoption of industrial practices during the 1970s and 1980s. Nationally, the impact of this institution's attempt to technify the coffee sector has not succeeded as other institutional forces have in other countries. Still, in areas like eastern Chiapas, the modernisation program carried out by INMECAFE coincided spatially with areas of recent social upheaval.

Colombia presents a good example of the degree to which a well-organised national institution can implement a policy it deems necessary. A key institution is the Colombian National Coffee Growers' Federation, known as "FEDERACAFE." As part of a strategy to increase production and to revamp the Colombian coffee industry, FEDERACAFE advocated the technification of production. The experience of coffee growers in the community of El Palmar illustrates the organisation's role in the process. Located in the southwestern Andes in Colombia's department of Valle de Cauca, growers in El Palmar received the full attention of the Federation's efforts designed to allow Colombia to take advantage of the high international coffee prices that were hiked up after the Brazilian coffee frosts of 1973 and 1975. The technification efforts included the use of the high-yielding variety *caturra,* the reduction of shade cover, and the intensification of agrochemical use.

El Palmar is only one example. Whereas twenty-five years ago, the majority of Colombia's area and production was associated with traditional systems, today 755,000 hectares of the country's total 1,104,000 hectares (2,728,000 acres) of coffee (68 percent) are technified. With modern farms capable of producing more coffee per unit area, the actual production on the technified area accounts for 86 percent of Colombia's total coffee produced. In years of high international coffee prices, Colombia uses upward of 400,000 metric tons of chemical fertilisers in coffee production.

Aside from national institutions, international aid programs have also played a part in the industrial transformation of the coffee sector in northern Latin America. Most notable is the United States Agency for International Development (USAID), which, for a 15-20 year period beginning in the early to mid-1970s, instituted a series of projects aimed at increasing production for the small coffee producer in several countries of Central America and the Caribbean. USAID's strategy for bringing more capital to the rural sector has involved technology transfer to small growers.

The technology being transferred has normally entailed a more industrial approach to production, including shade reduction and heavy chemical inputs. During this coffee technification period, the total cost of USAID projects in the region totalled about $80 million. There are still USAID projects aimed at coffee growers in El Salvador, Haiti and Guatemala. Some recent and pending USAID funding will be directed to environmentally sound coffee production via the development of market connections in some of the region's "smaller economies." This new hemispheric-scale project with $10 million to spend on free market support mechanisms will supposedly target small coffee producers in small coffee producing countries. One of the emphases will be fostering organic coffee markets for producers in Latin America and the Caribbean. The agency's El Salvador project has a strong environmental component, with organic coffee production playing a strategic role in capturing specialised markets.

Changing Coffee Production Technology

The changes occurring in coffee production technology are part and parcel of the general trend in world agriculture characterised by a progression toward evermore intensive practices. Since agriculture's beginnings, and especially since the advent of the industrial revolution, a relentless march within agriculture has continually refined, reshaped and sometimes remade farming and crop cultivation in the image of industry. Parts of the production process in a host of crops around the world have changed significantly over the past two or three centuries, giving rise to completely new cultivars, hybrid varieties, labour regimes, chemical inputs, and, ultimately, foods upon our tables.

These changes, as well as many others, increase the efficiency and volume of production, similar to parallel technological introductions in the industrial sector. They are part of a concept of "modernisation" that has proceeded in agriculture without sufficient regard to environmental consequences. Topsoil loss on erosion-prone croplands, pesticide poisoning of workers and the groundwater supply, huge increases in energy costs to bring one bushel of corn to the grain exchange, and an ever-increasing number of resistant insect pests appearing on the scene are just some of the problems facing industrial agriculture at the close of the twentieth century.

For the issue at hand — the transformation of coffee production — perhaps our task is to redefine "modern." While modern generally refers to the latest version or method associated with some phenomenon, the concept is incomplete without a component that reflects today's environmental challenges. This implies a commitment to using and understanding the best knowledge available to get something done in such a way as to minimise the impact upon the land. There are plenty of industrialised aspects of agriculture that can be used wisely in production. At the same time, there is a mountain of information that can be gleaned from generations of producers who have made it their everyday business to produce in such a way as to minimise risk and protect the land they use.

For coffee production, there are generations of knowledge collectively housed in the cultures of northern Latin America based on lifetimes of practical applied work. The traditional coffee system and the knowledge base associated with it is a veritable library of successful cultural practices. It is these practices that should be examined and tested for their present-day applicability and compatibility with more recent techniques. Being modern, in short, should incorporate the best knowledge from whatever system — old or new, "folk" or "scientific" — that preserves the productive base upon which longterm production depends. Hence, a truly "modern" production system can be regarded as one that benefits the grower with sustained yields and lowered costs over the long run, while at the same time maintaining or enhancing biodiversity, as well as protecting the land from erosion, chemical contamination, and the inhibition of natural nutrient cycling.

Distinguishing characteristics of traditional and intensified coffee production technologies

	Traditional	*Intensified*
Varieties used:	arabiga (tipica), borbón (bourbon), maragogipe	caturra, catuaí, Colombia (in Colombia), Garnica (in Mexico), catimor
Size (meters):	tall (3-5m)	short (2-3 m)
Shade:	moderate to heavy, covering 60% to 90% of ground area	none to moderate, covering up to 50% of ground area
Shade trees used:	tall (25 m) natural forest species, fruit trees, bananas	short (5-8 m), selected leguminous species (heavily pruned)
Density of coffee plants: (number per hectare)	1000 to 2000	3000 to 7000, with some areas up to
	10,000	
Years until first harvest:	4 to 6	3 to 4
Plantation life span:	30 years (and more)	12 to 15
Agrochemical use:	none to low	high
Pruning:	sometimes not pruned at all; otherwise, individualized treatment of plants	standardized stumping back* after first or second year of full production (*soqueo* or *recepa*)
Labour requirements:	seasonal for harvest and pruning	year-round maintenance with higher demands at harvest

Decisions about what technologies are ultimately used, as well as how best to insert oneself into the increasingly complex international marketplace, must obviously be made by those producing coffee. There is an urgent need to create the conditions in which social, political and economic structures allow such decisions, and provide growers with a wide range of knowledge upon which to base these decisions.

19

Organic Coffee, Protocols, Standards and Registration Procedures

Organic coffee is coffee grown completely free of synthetic chemicals. The land must have been free of synthetic pesticides and fertilisers for the past three years as a pre-requisite for registration. Burnett 1998, states that Organic coffee is the fastest growing segment of the US$2.5 billion Speciality Coffee Market, although it accounts for only about 5% of the market in the United States. He points out that certified organic coffee farmers earn 15-20% more for their beans than non-organic coffee farmers. Also, he points out that Organic Coffee is not necessarily any purer than coffee grown with pesticides as the pesticides are destroyed in the roasting process, with perhaps the exception of DDT residues-although DDT is rarely used on coffee today. However, buying Organic coffee supports a system that is improving the lives and health of poor farmers and the environment and helping to provide more equity for such people.

Organic coffee promotes the use of many sustainable agricultural practices, which conserve and protect and often improve the environment. Organic coffee has been growing at a rate of 25% per year since 1993, and in the United States, sales are far larger than Sustainable Coffee or Fair Trade Coffee.

Organic coffee certification is an expensive process and annual inspections may cost thousands of dollars for coffee cooperatives. Organic coffee may or may not be more labour intensive, depending on the farming system used for its production. This is an issue we hope to explore more in the Round Table meeting, along with procedures required and the practical experience of organic certification of Organic coffee in East Timor.

There is considerable debate between various players on what is Sustainable coffee and what is Organic coffee. Adam Tietelbaum from Adam's Coffee, and members of the Organic Coffee Association (ORCA) established in 1998/99 in the United States, maintain that with regard to coffee, "If it is not organic it is not sustainable." This view is not necessarily shared by others such as the Sustainable Coffee Coalition, which describe Sustainable coffees as those coffees grown with low or preferably no synthetic chemical inputs into the system. Tietelbaum 1997 indicated that the above mission statement did not indicate which petro-chemicals qualify as 'low toxicity' and can be used on coffee, and that a whole new certification bureaucracy, with all its associated costs, is needed to certify Sustainable coffee.

Standards and Protocols

A number of organisations have published standards, both general and specific for the certification of Organic coffee production. The Regulating Council on Organic Agriculture of the European Union requires that organic product inspection bodies conform to guideline EN 45011 and ISO Guideline 65. Many certified organisations for organic products are members of the International Federation of Organic Agriculture Movements (IFOAM), for example, BFA in Australia.

As an example, Biological Farmers of Australia Cooperative Ltd., (BFA) is a member of IFOAM, and BFA is nationally regulated by the Australian Quarantine Inspection Service (AQUIS), of the National Government of Australia. BFA operates a Total Quality Management System, accredited to ISO 9002, while conforming to ISO 65 and IFOAM guidelines. The regulation of organic farming and processing in Australia is based on a partnership approach between AQUIS and the organic industry through various AQUIS approved organic certifiers such as BFA.

Under this partnership the approach industry is responsible for setting organic standards and delivering services directly to exporters and operators while AQUIS is ultimately responsible for enforcement of industry standards. To do this AQUIS approves the individual organic certifier bodies as well as regularly auditing the performance of those bodies to ensure they are properly carrying out their certification and inspection functions. The AQUIS-approved organic certifiers inspect organic producers, processors, exporters and products and issue Organic Produce Certificates to allow the export of complying products.

Improving the Quality and Preventing Mould Growth

Good quality coffee receives bigger payments. In order to protect your revenue from coffee, it is essential that you provide only the best quality commodity available.

Use Good Agricultural Practices

— Cherries that have dried on the tree and those that

have fallen to the ground are known to be susceptible to mould growth and therefore these should not be picked.

- Process cherries as quickly as possible. Avoid storage of cherries, especially ripe and over-ripe ones, as any period of storage increases the likelihood of mould growth.
- Do not dry on bare soil. Mould spores from previous lots are known to remain on the ground and this could result in clean cherries being contaminated during drying.
- Protect cherries during drying from rain and night dew.
- Avoid re-wetting of drying cherries.
- Protect dried cherries from the moisture and rain.
- Clean coffee from all husk material-more than 90% of mould comes from husks in sun dried cherries.
- Remove as many defects (husks, un-hulled cherries or mouldy beans) as possible.
- Keep cleaned dried beans separate from discarded material.

During processing

- Site processing plant in good conditions.
- Provide good drying facility-turn regularly.
- Prevent recontamination by avoiding contact with dust, husks, and dirty bags.
- Do not store cleaned, dry, graded coffee near rejects and husks as cross-contamination may occur.
- Keep moisture content as uniform and as low as possible-below 13%.

During transport and storage

— Cover bags during transport and storage to prevent re-wetting.

— Load and unload trucks or containers only in dry weather or under cover.

— Do not use damaged containers and prevent water leaks.

— Make sure that pallets or wooden floors of trucks and containers are dry.

— Store coffee in well-ventilated and leak-proof warehouses. Store away from the walls.

— Cover bags/beans in container with waterproof or water-absorbent cover to prevent re-wetting from condensation.

— Provide good quality control tests and ensure that they are adhered to, especially to check for moisture and defects.

Global Perspectives in Quality Improvement

Each year 6 million tons of coffee are produced, more than 80% to be consumed as roasted and ground (R&G) brews, while a little less than 20% goes into the production of soluble coffee. Within the roasted and ground coffee are different types of brews, ranging from the crystal clear filter coffee to thick and foamy espresso, 3-4% of total consumption, but fast-growing worldwide. Within the soluble coffees we find liquid coffees, becoming important particularly in the Far East, and powder coffees, picking up a 'share of the throat' especially in the tea-drinking countries of the world. The required quality depends on the use, which does not mean that any coffee should finally find a buyer.

Unfortunately this is sometimes the case and poor quality coffee finds its way onto the market.

Quality at Production

Quality of green coffee depends on:

- *Climate, soil, species and breeding characteristics:* In breeding programs, beside resistance to diseases, productivity, and morphological traits, cup quality must always be considered.
- *Harvest:* The most important parameters are ripeness of cherry and time to processing. The best coffee may give an astringent or impure cup, if harvested immature or if kept too long before processing.
- Processing to green beans, simplified here as:
  - dry-processing with development of rich body and aroma;
  - wet-processing, for fine aroma and acidity.

When climatic conditions allow it, dry processing of Arabica coffee adds to the body of the liquor, a characteristic appreciated by the fast growing espresso segment. All downstream operations can at best only maintain the liquoring qualities attained by green coffee, never improve or correct them.

Quality for the Roaster

Quality evaluation for the roaster requires:

- reliability of supply;
- uniform low moisture and agreed defect count;
- regular roasting characteristics;
- cup quality.

Quality for the Consumer

Although consumers generally do not possess a refined vocabulary to explain if their likes or dislikes for a particular cup of coffee, which is often either 'bitter' or 'good', their consumption patterns are strongly influenced by taste and smell-top quality coffees always find consumers who can afford them.

In the recent years, environmental (green issues), such as Organic, Fair Trade or Sustainable coffees have also become criteria of choice for the consumers. These new quality criteria can bring benefits for the planters who receive a guaranteed minimum price, or a bonus for above-standard quality and advice on quality control and market needs. Roasters also benefit by ensuring that the farmers produce coffee according to the required standards, and may sell their product under a special label.

Quality for the Regulator

In the interest of consumer protection, the regulator acts from purity and from safety considerations.

Quality Parameters for the food technologist

Many hundreds of the compounds formed at roasting by the chemical interaction between the carbohydrates, chlorogenic acids, amino acids and other reactive compounds present in the green bean have been identified. Differences are minor, mainly quantitative, and the complete chemical profile, cannot be used to explain why one coffee gives a better cup than another.

Aroma

The most important parameter in the appreciation of quality is the organoleptic quality of the cup, mainly due to the volatile substances present, accounting for no more than 0.1% of the total, while non-volatile components can only explain acidity and bitterness.

Among the many hundreds of components discovered in the aroma complex, the active smelling compounds have now been identified by sniffing all the components coming out of the column outlet of a gas-chromatograph. The most intense smelling components-the majority already present in green beans, may then be identified by successive dilutions of the aroma until only a few can still be detected (Table 1).

Table 1. Main odorants in a coffee brew (µg/l)

Aroma	*Arabica*	*Robusta*	*Threshold in water*
(E)-β-Damascenone	1.3	1.5	0.00075
3-Mercapto-3-methylbutylformate	5.5	1.5	0.0035
2-Furfurylthiol	19.1	39.0	0.01
Methanethiol	210	600	0.2
3-Methylbutanal	550	925	0.35
Methylpropanal	800	1350	0.7

The same technique helps in identifying the off-flavours that, if present even in extremely minute amounts, spoil cup quality (Table 2).

Table 2. Main off-flavours of coffee

Off-flavour	*Responsible chemical(s)*	*Threshold in water (µg/l)*
Medicinal, 'rioy'	2,4-6-Trichloroanisole	0.001
Earthy, Robusta	(-)2-Methylisoborneol	0.0025
Musty	Geosmin	0.005
Fruity, rotten, stinker	Ethylesters of 2- and 3-methylbutanoic, and cyclohexanoic acids	5,000-10,000

All these compounds are probably of microbiological origin, and are already present in the green bean.

A technique which proves very useful in the evaluation of new processes, particularly in soluble coffee manufacture, indicates the flavour evaluation by a trained test panel in the form of a star diagram, where notes like burnt, cooked, fruity are profiled semi-quantitatively.

Quality Parameters for the Roaster

The European Contract for Coffee states that, *"all goods contracted for shall be of sound merchantable quality..."* A new version of the contract, including a reference to excessive moisture as not being in conformity with the quality requirements, is under study by the European Coffee Trade Federation. Both analytical (moisture, defect count) and organoleptic (taste testing) criteria are available and are used by the roasters for the choice of the green coffee qualities they use in their blends:

— Speciality coffee roasters, espresso in particular, need special qualities (e.g., body from top dry-processed Arabicas, absence of immature beans, which make the cup astringent and metallic);

— Major roasters need large quantities of reliable constant good quality coffees, particularly Robustas.

Measurement of Moisture

Several different ISO standards are available for the measurement of moisture. They do not all give the same results, thus contradicting the idea of standardisation. In the last meeting of ISO TC34 (Agriculture) SC 15 (Coffee), it has been agreed:

— to select among the methods already available the most suitable to be used as reference;

— to standardise a rapid method, based on devices, such as Sinar, Dickey-Jones or Dole, which have become common in the trade.

Defect Count

Many different grading systems are used in the trade of coffee and at the last meeting of ISO TC34 (Agriculture) SC 15 (Coffee) a revision of ISO 10470, simplifying it, was proposed:

— by considering together both dry and wet processed Arabicas and Robustas;

— by classifying defects according to their incidence (1) or (0) on the organoleptic profile of the cup, and on the economic aspects;

— by indicating if the defect is serious (s), so that it could be applied generally in the trade.

The weight of each defect is also being reassessed in the light of new information available, and the following defects are now considered as serious:

— Organoleptically-beans that are black, partly black; dark brown, amber, with foxy silverskin, sour, stinker, spotted bean; pergamino, bean in pergamino; pod, husk fragment; mouldy bean.

— Commercially-bean fragments, bean in parchment; large, medium, small sticks; large, medium, small stones; soil agglomerates; foreign matter.

That new criteria must be considered in the evaluation of defects has also been agreed by the London International Financial Futures and Optional Exchange (LIFFE) classification of Robustas, stating that "... coffee is not tenerable if:

— it has more than 450 defects per 500 g;

— it is unsound;

— it contains more than 10% passing through a round screen 12;

— in respect to a lot graded, it has more than 5 fully mouldy or 10 partially mouldy beans in combination thereof, such that the total exceeds the equivalent of 5 fully mouldy beans per 500g."

Liquoring

The techniques of organoleptic evaluation, useful in the development of new processes, would be too cumbersome for the roaster wishing to ensure routinely the wholesomeness of the raw material used. A simple and clear vocabulary is in general sufficient to a trained expert panel in the day-to-day liquoring routine:

For Arabica, flavour can be defined as the sum of *aroma* plus *acidity* plus *body*, where

— flavour is the taste of a sound, clean Arabica coffee, not the level of roastiness;

— aroma is the smell of a sound, clean, freshly brewed Arabica coffee;

— acidity is a sharp and pleasing taste as opposed to a sour taste, which may indicate signs of fermentation. Acidity is best appreciated in a low roasted filter coffee;

— body is the viscosity, fullness and weight in the mouth, ranging from thin and watery to thick and heavy. It is an important characteristic particularly in espresso coffee, where it is associated with a good body, as shown by the comparison of a correctly prepared cup with a poorly prepared one when using the same blend.

The panel must also recognise a few undesirable flavours and all off-flavours, such as:

- — undesirable flavours of Arabica-green/grassy, cereal/woody/papery, baggy;
- — off-flavours of Arabica-chemical/medicinal, hard/ metallic, earthy, fermented, mouldy/musty.

The flavour of a Robusta must be neutral/bitter, and devoid of woody and rubbery notes. Undesirable/off-flavours of Robusta are: green/grassy, chemical/ medicinal, and earthy, fermented or mouldy/musty.

The organoleptic profile is usually obtained by one of two techniques:

- — by oral agreement within an expert tasting panel after open discussion; this technique is the most effective if the aim of the panel is to identify the top quality coffees;
- — by the average of independent results (either blind or by comparison with a reference), of members from the tasting panel. This technique is useful for maintaining a good constant average quality in industrial production.

Quality Parameters

Purity

Lack of physical criteria, such as defect counting, has hindered the objective evaluation of soluble coffee, particularly when imported from producer countries, where no control of stocks could be easily achieved, and only taste testing could give some indication of the quality of the product.

Findings, showing that there was a precise carbohydrate fingerprint for pure soluble coffee, led to the establishment of an ISO analytical standard, and to national Codes of practice in the United Kingdom and France, which, by indicating maximum acceptable levels of certain carbohydrates, have helped in reducing the import of adulterated products into the European Union (Table 3).

Table 3. Tolerable control limits for carbohydrates in soluble coffee

Indicator Carbohydrate	*Maximum Content in Pure Coffee (%)*	*Control Limit (%)*
Total glucose	2.1	2.6
Total Xylose	0.4	0.6
Free Fructose	0.6	1.0

The question of soluble coffee purity has become particularly important in Eastern Europe after the opening of the markets in the 1990s, and must still be solved, particularly after a link between relatively high contents of Ochratoxin A, and adulteration has been found.

Safety

The possible presence of contaminants, such as pesticides, polycyclic hydrocarbons formed at roasting, or paraffins from the coffee bags, have at various time alerted authorities until analytical data were made available indicating that the safety problem had been solved. Mycotoxins, Ochratoxin A in particular, may be formed during cherry processing, storage or transport of the beans, if moisture is uncontrolled, and will still be present in the cup, even if it is partially destroyed during roasting. A multicenter project is active to solve this new challenge.

20

Coffee Preparation

Coffee Preparation Rules

Great tasting coffee depends on proper preparation. Some simple rules:

1. Clean the coffee brewer, pot, machine, thermos, whatever.
2. Use good tasting water.
3. Use the correct coffee grind size for the brewing device.
4. Use the recommend amount of coffee for the device.
5. Serve fresh.

Most of these rules are obvious, but the one most commonly disregarded is cleaning the equipment. Nothing but nothing makes good coffee taste bad like a dirty brewer or thermos. Plastic brewers are really bad. The plastic is porus and can hold many elements which can and do make the coffee taste bad. At best, taste bitter and rancid. Not to mention possible disease. Clean the equipment regularly.

Preparation

Coffee is very versatile. Coffee stands on its own, can act as a food, and is commonly used as flavouring.

Filtered Coffee

Most the world is poor and can not afford the fancy machines citizens in developed countries can. And, many people in the developed countries are poor too. This method is extremely common but most the people with Internet web may not believe it. It is a home style of making coffee. Pour the brew through a sock. A filter for the purpose of filtering coffee will work better, but a cotton sock will work.

Coffee brewer

Using medium size grind, put the grind in a basket type filter and put the filter in a pot, bowl, or large glass, pour in hot water, or boil the water in the pot. Remove the filter with the grounds and wa-la, you are ready to drink the brew.

For best results, wash the pot at regularly and clean the filter regularly. Once a week or at least once a month, boil a pot of vinegar in the pot and through out; then boil a pot of baking soda in the pot and through out; wash thoroughly. There are special cleaners for this purpose too. Warning: do not use general metal cleaners, only use those which are made for this purpose. Believe this or not, you can use an old sock. Be sure to wash the sock before use. It should be white and not furry. Cotton works well.

If you are boiling the water in a pot with the grinds, be sure not to over boil. Many of the compounds will begin to break down and the taste will get very bitter.

This process works well on a stove top, in a microwave oven, and over a camp fire.

Drip System

The worlds next most common method of making coffee is with a drip system. They are found in commercial use in most restaurants. They are also very common in households.

The process is put the ground coffee in a container on top of the machine. If the machine does not have a hot water maker associated with it which usually pumps the water to the grinds, poor the hot water into the top container which holds the grinds. The water then saturates the coffee grinds and drips out into another container.

These come in many different styles. Two distinct differences are those which heat the water in the machine and those which hot water is pored into them. Another distinct variant are those which keep the brewed coffee hot by a hot plate where the brew container is and those that do not. Some use just a metal filter and others use a paper filter. Additionally, they can be made of metal or plastic. These are fairly simple and cheap. But, not a cheap as an old sock.

Drip brewer

Be sure to use the right size grind for the device's filter. Typically, metal filters require a larger grind size, and paper filters use a smaller grind size.

For best results, wash the ground container regularly and clean the filter if metal regularly. If metal, once a week or at least once a month, put a pot of vinegar water through the system; then pot of baking soda and water through the system; wash thoroughly. There are special cleaners for this purpose too. Warning: do not use general metal cleaners, only use those which are made for this purpose. Plastic is porus and becomes contaminated quickly. Cheap cleaning methods do not work on plastic and can wreck the plastic. It's best to replace the plastic device regularly.

Percolator

The next most common device for brewing coffee is the percolator. It is also considered the poorest form of brewing coffee because it reboils the brew constantly which tends to make the brew bitter. The device is very common in households but not common in commercial establishments.

Percolator

Generally these machines are self contained and electric. However, there are devices using this method which utilize an external heating source.

The device works by putting grinds in a container on top of the system. A special heat concentrator and collector at the bottom of the device allows water to boil on the bottom of the system and a tub funnels the bubbling water to the top where it is dispersed over the grinds. The hot water soaks down through the grinds and drips back into the main container. The system circulates the water regularly through the grinds until done.

Generally, these devices are metal but there are those made of glass.

For best results, wash the pot at regularly and clean the filter regularly. Once a week or at least once a month, boil a pot of vinegar in the pot and through out; then boil a pot of baking soda in the pot and through out; wash thoroughly. There are special cleaners for this purpose too. Warning: do not use general metal cleaners, only use those which are made for this purpose. Unfortunately, the percolator grind holder does not get very clean with this method so an added soaking step should be used. After each boiling cleaning step, pull out the peculator stem and allow the filter holder to sit at the bottom of the device and soak, then go to the next step.

Coffee Press

The Plunger or Coffee Press was invented in France in the 1850's. Originally a one stage device which pushed the coffee down into the brew. There were several problems with this device so a two stage plunger has been designed to overcome most of these objections.

The two stage plunger allows coffee to be made in a microwave oven but any hot water will do. The glass/plastic body is placed in the microwave oven without the metal filter parts and the water boiled. The oven will indirectly heat the plunger body and keep the brewing temperature around the desired level. The coffee is spooned onto the water and floats dry. With both filters together press down until they are below the water level mark. Then, pressing the lid down, the lower filter is forced through the grounds and causes the coffee to stir and steep. Pressing the knob down after the correct brewing time, coffee grinds are fully pushed to the bottom. The knob must be in down position when lid is removed from pot.

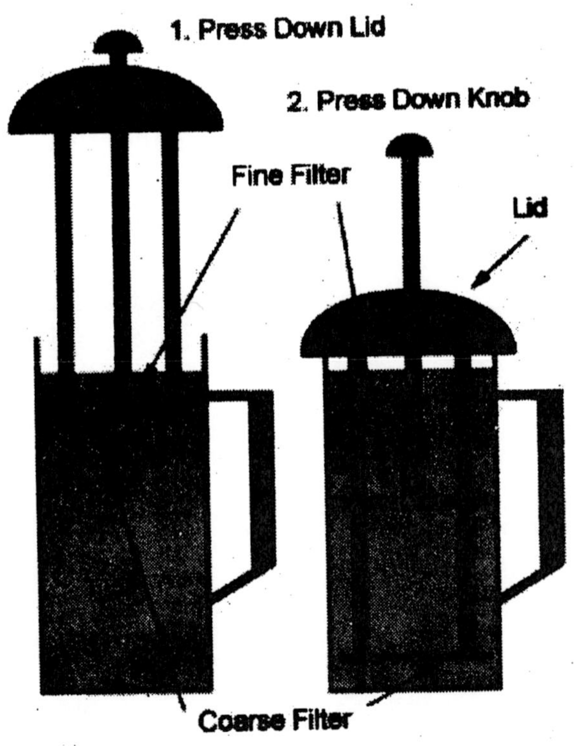

Sieve brewer

The result is coffee brewed with hotter water and held at hotter temperature. Provided the brewing time is correct, results in the double plunger will be better than a single plunger. This is due to the hotter water, the improved steeping and stirring, the better brewing temperature and the hotter serving temperature.

Espresso and Expresso

Espresso is the most complex way of making ever developed. Literally in Italian, expresso means coffee made on demand. It is universally considered the best way of making coffee. Generally speaking, the degree of roast is City, Full-city, Italian, French, High, Dark, or Spanish. There is a definite point in the espresso action where the degree of roast yields the best results. Lighter roasts as Brown, American, Medium, Regular, Cinnamon, and Light tend to be very poor. Dark roasts as Napoleon, Heavy, or Dark French are ok but extraction is lower.

Expresso brewer

We are not going to tell you how to use a steam driven machine because we don't want to get sued. It both burns people and blows up. We will warn you to check for bad fittings which should be replaced as they can fail with disastrous consequences. And, don't over heat the steam.

What we are going to tell you is how to use an expresso machine. An expresso machine will make a cup of espresso because it is what the machine is designed to do. The expresso machine is designed to make a cup of espresso easily and more or less fool proof.

Expresso brewer

There are still some human touches to making a good cup of expresso. One real good human touch is the choice of degree of roast. The above degree of roast is exactly the same for an expresso machine as it is for the common espresso.

If one is using the cheaper machines which do not automatically pack the grind and pump the water then it is important to note how to do it. In filling the grind container, fill to the top and depress with the tamper. Press firmly, but not hard. One may also tamp the grinds into place. Too tight or too loose does not allow the water to pass through at the right rate. In dispensing the water from a lever action, take from 20 to 30 seconds to complete the process. Don't dispense water twice just to get more fluid. It defeats the purpose.

To froth the milk, put the milk in a container, hold container under the steam vent, then slowly turn the steam knob until steam permeates the milk. Turn off the steam vent, then remove the milk.

This machine must be cleaned thoroughly every day. The steam vent must be cleaned just after use. Also, because of the way the machines are made, it is wise to use water with low mineral salts so build up does not occur.

Sweet Espresso

Add sugar to the espresso. Because espresso is usually a small cup of coffee, lumps of sugar are not all that good. They work, but it is better to use granules.

A very common sweetener is honey.

Espresso and Milk

Theoretically, the story behind espresso and milk is to trap the aroma. Espresso has much aromatic properties which vaporize quickly. So, a thin layer of milk is used to trap the aromatic properties in the cup.

Froth

Froth is the process done to milk to make it foam. All espresso and most expresso machines have a special tub to make steam which is used to make froth. Typically, the froth is made with a thick milk as opposed to skim milk. The milk is in a container with a spout for controlled pouring. The container is held under the tub and a valve is turned which allows steam to froth the milk. Basically, it is making hot foamy milk.

The milk is carefully poured into the espresso in such a manner not to sink or mix with the coffee.

Espresso Macchiato

Espresso with just a stain of milk. A very small amount of milk just to stain the coffee.

Latte Macchiato

This is the opposite of staining the espresso with the milk. This is where the milk is added first and the espresso is used to stain the milk.

Caffe Mocha

One-third espresso, one-third hot chocolate and one-third frothed milk served in a large class. This is not to be confused with Mocha-Java coffee beans. It doesn't mean it comes from Mocha either.

Con Panna

Coffee with whipped cream. May be espresso but sometimes it can be any cup of coffee.

Double

Double Cappuccino, double Caffe Latte, or double Caffe Mocha means a double helping in espresso or otherwise of everything. However, some people think it means double the coffee but not the fixings. Oh well.

Extra Strength

Extra strength means more double the coffee but not the water. This is a condition where one would use more coffee grinds than normally called for.

Iced Espresso

Usually a double espresso poured over lots of crushed ice in a fancy glass. Sometimes topped with whipped cream.

Iced Cappuccino

Espresso poured over crushed ice, topped with equal part of cold milk, and topped with cold frothed or whipped milk.

Espresso Granita

Italian granitas involve freezing strong, unsweetened espresso, crushing it, and serving it in a parfait glass or sundae dish topped with sweetened whipped cream.

Cappuccino

Typically, a 1.5 ounce of espresso, topped by hot milk and foam. Ideally, cappuccino consists of on-third espresso, one-third hot milk, and one-third foam. May or may not be drunk with sugar.

Latte

Many times referred to as Caffe Latte or cafe au lait, or cafe con leche. Simply the same as cappuccino with much more milk. Additionally, it is mixed by pouring the milk and the coffee from opposites sides of the cup. Although traditionally made with espresso, it dilutes the coffee so much that it is now made with standard brews.

Mocha Latte

— 1/4 espresso,

— 1/4 hot chocolate,

— 1/4 hot milk,

— 1/4 foam.

Made generally the same as Latte. Pour the espresso, hot chocolate, and milk together, then top with foam.

Irish Coffee

— 1 teaspoon sugar

— 1 or 2 tablespoons Irish whiskey

— regular black coffee

— optional cream or lightly whipped cream.

Schmidty's Irish Coffee

— 2 teaspoons granulated sugar.

— 2 teaspoons Irish whiskey.

— 2/3 cup of hot French roast coffee.

— Heavy cream lightly wipped as garnish.

Place surgar and wiskey in a cup then mix till surgar has gone into solution. Pour in coffee then carefully pour cream on top. Do not mix in cream.

Frappe

— 1-2 teaspoons of instant coffee.

— 2-3 teaspoons of sugar.

— milk with ice-cubes.

Frappe coffee is widely consumed in parts of Europe and Latin America. It is made with cold espresso. It is prepared in most places by shaking into a shaker with instant coffee, sugar, milk and ice-cubes. It is served in a long glass with ice and a straw. One important thing is the thick froth on top of the glass.

Frappe Alexander

— 2 cups chilled double strength coffee.

— 3 tablespoons granulated sugar.

— 1/4 cup brandy.

— 1/4 cup heavy cream.

— 1/4 cup cream de cacao.

— Ice cubes.

— Nutmeg for garnish.

Combine ingredients in blender till frothy. Pour in classes. Typically serves 4. Top with nutmeg.

Coffee Royal

A regular cup of coffee with whiskey or bourbon.

Turkish Coffee

Turkish Coffee requires special grind. It is hard to come by but can be purchased at a good specialty coffee outlet. The grind is extra fine. It is a fine powder.

Put a table spoon or more, in a coffee cup and nuke in the microwave oven. Take it out and drink it.

It can be made on a stove too. However, because a microwave makes water hotter than a stove, microwave ovens are starting to be used for this purpose.

There is no need to filter the grinds out because they may settle out.

Old-fashioned Coffee Soda

— 3 cups chilled double strength coffee

— 1 tablespoon superfine grind sugar

— 1 cup of heavy cream; or half&half.

— 1/3 cup of chilled club soda.

Wipped cream and marachino cherries for garnish.

Ciudad Cooler

— 4 cups extra strength coffee.

— 2 cinnamon sticks.

— 4 whole cloves.

— 3 whole allspice.

— 1/2 cup of heavy wipping cream.

Mix the coffee, cinnamon, cloves, and allspice in a large pot and steep for 30 minutes. Fill four tall galsses with cracked ice. Strain the coffee through a filter and pour int each glass.

Turkish Cola Float

— 1 part chilled extra strength coffee.

— 1 part dark coffee ice cream.

— 1 part cola.

Scoop the ice cream into a glass, pour in coffee, and carefully pour in cola. Cola and ice cream fiss, so lit takes a little care in mixing this drink.

Cappuccino Borgia

— 1/4 peeled orange.

— 1/4 cup of coolled espresso.

— 1 1/2 cup of chocolate ice cream

— 6 tablespoons orange juice.

— 1/4 cup whole milk.

Whipped cream, grated orange, and chocolate bocca beans for garnish.

Mix the espresso, orange, ice cream, orange juice, and milk in a blender. Blend until smooth. Pour into glass and garnish.

Calypso Cooler

— 1 cup chilled extra strenght coffee.

— 2 small ripe bananas.

— 3 cups coffee ice cream.

— 4 tablespoons rum.

— Ground cinnamon for garnish.

Puree bananas and coffee in blender, then add ice cream and rum, blend till thick. Pour in glass and sprinkle with cinnamon.

Angostura Cooler

— 2 cups chilled extra strength coffee.

— 2 cups vanilla ice cream. 1 tablespoon Angostura Bitters.

— Whipped cream and chocolate curls for garnish.

Blend ingredients, pour in 4 tall glasses, fill with whipped cream, and top with chocolate curls.

Supreme Bean Grog

— 1/8 teaspoon ground cinnamon.

— 1/8 teaspoon ground allspice.

— 1/8 teaspoon ground nutmeg.

— 1/8 teaspoon ground cloves.

— 2 tablespoons of butter.

— 2 cups of packed brown sugar.

— 1 cup of rum.

— 2 cups of hot Italian-roast coffee.

— Grated orange for garnish.

Combine spices in small bowl and mix. Beat butter until light and add sugar slowly until blended. But 1/8 teaspoon is spices in cup, pour in 1/2 cup of rum, 1 cup butter/sugar mixture, and add 1/2 cup of coffee. Makes 4 one pint mugs.

Caffe Zabaglone

— 4 egg yolks.

— 1/4 cup granulated sugar.

— 1/8 teaspoon salt.

— 1/4 cup dry Marsala wine.

— 3/4 cup Italian roast coffee room temperature.

Beat the egg yolks, sugar, and salt till creamy pale yellow. Slowly stir in wine. Cook over a double boiler and wisk till mixture becomes a thick foam and soft mounds form. Fold in coffee until blended. Pour into 4 goblets. Serve with spoon.

New Orleans Coffee Eggnog

— 1/2 cup of eggnog.

— 1/2 cup of double strength New Orleans coffee (American roast).

— 1/2 cup of heavy cream.

— 1/2 cup of bourbon.

Nutmeg for garnish. Whip eggnog, coffee, and cream till thick. Fold in bourbon. Garnish. Serves 2.

Royal Coffee Punch

— 4 cups extra strenght coffee cool.

— 1 cup powdered sugar.

— 1 cup brandy.

— 1 megnum champagne, chilled.

Combine in a large pitcher or bowl. Mix gently. Serves 16-12.

Sambuca Mosca

— 2 shots Sambuca.

— 5 roasted coffee beans.

Pour Sambuca in a liqueur glass, drop on beans. Light with match. Serve imediately.

Cardamom Kaffe

— 1 Cardamom pods, cracked and seeded.

— 1/8 cup cognac.

— 2 teaspoons curacao.

— 1 teaspoon sugar.

— 1/2 cups extra strength coffee.

Combine Cardamom pods, cognac, curacao, and sugar in a cup. Heat gently or NUKE in microwave oven for about 10 seconds and light. Serve with a siren. Allow guest to put out burning mixture with the 1/2 cup of coffee.

Chili and Beans

— 1 cup of dark roasted coffee beans.

— 1 can of chili and beans.

— 1 cup Vodka.Sour cream and cheddar cheese for garnish.

Soak roasted coffee beans in Vodka over night. Pour off the Vodka into two glasses. Mix coffee beans with can of chilli and beans then simmer, stir occansionally. Serve in 2 bowls with Vodka.

Coffee Sauce

— 1 cup granulated sugar.

— 2 cups extra strenght coffee.

— 2 tablespoons cornstarch.

— 2 tablespoons butter.

— 1/2 teaspoon salt.

Melt sugar slowly in heavy skillet till amber color. Slowly add 1 1/2 cup of coffee stirring constantly. Steams heavily. In a small bowl, blend cornstarch and 1/2 cup of coffee, then pour sugar mixture and cook till thickens. Remove from heat and stir in butter and salt till blended. Serve warm over any thing.

Coffee Candy

This is good. Melt semisweet chocolate in a pan and add sugar. Depending on taste, add some cream. Put in some fine coffee grinds. Typically, Turkish Coffee grinds. Pour into a candy bar container. Allow to harden.

Coffee Cinder

— 1 cup of over roasted Robusta coffee beans.

— 6 ounces of milk chocolate.

Melt chocolate. Coat a small chunk ice cube container with some of the chocolate. Put 3 to 5 beans in ice cube container. Pour remaining chocolate over beans. Allow to cool or chill.

Bibliography

Carvalho, A., "Principles and Practice of Coffee Plant Breeding for Productivity and Quality Factors: Coffea arabica", *Coffee: Agronomy*, Ed. R.J. Clarke. New York: Elsevier Applied Science, 1988.

Davids, K., *Home Coffee Roasting: Romance and Revival*, New York: St. Martin's Griffin, 1996.

_______________., *Coffee: A Guide to Buying, Brewing and Enjoying*, Santa Rosa: 101 Productions, 1991.

deGraaff, J., *The Economics of Coffee*, Pudoc, Wageningen, Netherlands, 1986.

Dicum, Gregory and Luttinger, Nina, *The Coffee Book*, New Press, 1999.

Illy, A. and Viani, R., *Espresso Coffee: The Chemistry of Quality*, San Diego: Academic P, 1995.

Illy, F. and Illy, R., *Dal Caffe al Espresso*, Milano: A. Mondadoni. In Italian, 1989.

Kamau. I. N.; "Mechanical Drying of Arabica Coffee in Kenya", *In "Kenya Coffee."* Vol 45. No 537. Dec 1980.

Knox, K., *Coffee Basics: A Quick and Easy Guide*, New York: John Wiley and Sons,1997.

Lingle, T., *The Basics of Cupping Coffee*, Second Edition. SCAA, 1993.

Mitchell, H. W., "Cultivation and Harvesting of the Arabica Coffee Tree", *Coffee: Agronomy*, Ed. R.J. Clarke. New York: Elsevier Applied Science, 1988.

Pendergrast, Mark, *Uncommon Grounds: The History of Coffee and How It Transformed Our World,* New York: Basic Books, 1999.

Rice, Paul D. and Jennifer McLean, *Sustainable Coffee at the Crossroads,* Washington, DC: Consumer's Choice Council, 1999.

Rice, R., "New technology and coffee production: examining landscape transformation and international aid in northern Latin America", Washington, DC: Smithsonian Migratory Bird Center, 1993.

Sherry, Thomas W., *Shade Coffee: A Good Brew Even in Small Doses,* The Auk, 2000.

Van de Vossen, *Kenya Coffee,* 1980.

Wrigley, Gordon, *Coffee,* New York: John Wiley and Sons, 1988.

Index